TRANSACTIONS

OF THE

AMERICAN PHILOSOPHICAL SOCIETY

HELD AT PHILADELPHIA
FOR PROMOTING USEFUL KNOWLEDGE

———

NEW SERIES—VOLUME XXXI, PART IV

DECEMBER, 1940

———

THE ROUTE OF COLUMBUS ALONG THE NORTH COAST OF HAITI, AND THE SITE OF NAVIDAD

SAMUEL ELIOT MORISON

———

PHILADELPHIA:

THE AMERICAN PHILOSOPHICAL SOCIETY

104 SOUTH FIFTH STREET

1940

LANCASTER PRESS, INC., LANCASTER, PA.

CONTENTS

LIST OF ILLUSTRATIONS AND MAPS

Photographs by the Author and Lieutenant Élie; Maps by
Dr. Erwin Raisz and Bertram Greene

THE ROUTE OF COLUMBUS ALONG THE NORTH COAST OF HAITI, AND THE SITE OF NAVIDAD *

SAMUEL ELIOT MORISON

Professor of History, Harvard University

1. INTRODUCTION

During the month of January, 1939, I made a personal reconnaissance of the north coast of Haiti, in order to ascertain the exact course of Christopher Columbus on his First Voyage, and to identify, if possible, the site of Navidad, the first settlement attempted by Europeans in the New World. Through the good offices of the Right Reverend Harry R. Carson, Bishop of Haiti, I was introduced to the chief of the Haitian army, Colonel André, who graciously placed at my disposal the coastguard cutter known as "Chaloupe No. 5." This comfortable and seaworthy little motor cruiser, commanded by Lieutenant Auguste Élie of the Garde de Haïti, and manned by an efficient crew of Haitian sailors, conducted me around Cape St. Nicolas and along the coast to Fort Liberté. There I was the guest of Mr. Robert L. Pettigrew, the principal sisal planter of Haiti, a man who takes a keen and intelligent interest in Columbian problems. Mr. Pettigrew and one of his colleagues in the sisal industry, Mr. Kenneth Crooks, conducted me by foot, car, mule, motor-launch and even steamer to and along the shores of the sound that extends from Cape Haitien to Yaquezi Point. These investigations, accompanied by an intensive study of the relevant documents and ancient maps, enabled me to arrive at a fairly definite conclusion as to the reef where the *Santa Maria* was wrecked, and the site of Navidad.

My quotations are largely translated by myself from the Las Casas abstract of Columbus's *Journal* of his First Voyage, which was first printed by Martín Fernandez de Navarrete in his *Colección de los Viages y Descubrimientos* (Madrid, 1825).[1] Navarrete and others have attempted to identify the various points mentioned and named by Columbus; but these identifications, being made from a study of maps without knowledge or observation of the coast, are for the most part worthless.[2] A coast line often looks entirely different as one approaches it by sea, from what one would suppose by studying the maps; and as Columbus had no maps, it appeared to me that the only way to ascertain his exact course was to follow the route indicated in his *Journal*.

The only liberty I have taken with the text is to introduce the word "Roman" wherever Columbus mentions "miles." I have done this because the Roman or Italian mile that

---

* All maps in this article are copyrighted by the author, to whom application should be made to reproduce in whole or in part.

[1] For the different texts and translations of the *Journal*, see my article, "Texts and Translations of the Journal of Columbus's First Voyage," in *Hispanic American Historical Review*, XIX (1939), 235–61. My translation was made from the Cesare de Lollis text in *Raccolta di Documenti e Studi Pubblicati dalla R. Commissione Colombiana* (Rome, 1892–94), Parte I, Vol. I.

[2] See especially Sir Clements R. Markham's map, reproduced in Filson Young, *Christopher Columbus* (3d ed., 1912), p. 236, which even reverses the order of places that the Admiral names in his *Journal*. Most of Columbus's biographers make no more effort to identify the places that he visited and named, than if his voyage had been one to the moon.

Columbus used was only 4,850 feet in length, and there were four of them to his league. The English (and now international) nautical mile, equivalent to one minute of latitude, measures 6,080 feet, and there are three of them to the English league. Consequently four of Columbus's miles equal, roughly, three nautical miles; and his league is equivalent to 3.18 nautical miles. It is, therefore, misleading to translate Columbus's *millia* as "mile" without prefixing "Roman." Whenever I use the word "mile" without qualifications, it should be understood as a nautical mile of 6,080 feet. It should be understood that the greater part of Las Casas's abstract Journal is in the third person, although occasionally he quotes the "exact words" of the Admiral.

The navigators of that day had no instrument—not even a chip log—for measuring distances at sea. They merely estimated the speed of their vessels through the water, and applied the elapsed time, which in turn was not measured by any instrument of precision, but by a half-hour glass. Any distance that could not be thus imperfectly measured by the time and estimated speed of a ship was a mere guess; and Columbus was rather fond of guessing the distance of a point when he took the bearing. He commonly overestimated the distance, and did not trouble to revise the log entry when later he found that the distance was less than he supposed; but his compass bearings were generally correct within a point.

The westerly variation of the compass on the North coast of Haiti is less than two degrees in 1939; in 1500, according to the diagram printed by Van Bemmelen,[3] it was even less; and this is borne out by Columbus's own sketch map of Haiti, which is oriented to true north. As there are $11\frac{1}{4}$ degrees to a compass point, a variation of two degrees is negligible in plotting courses, and I have, therefore, used the compass-rose in the current charts for my bearings, as varying only a small fraction of a point from the Admiral's compass.

The following table shows the Spanish names for compass points as found in Columbus's Journal, with the modern English equivalents, American abbreviations, and the equivalent values in degrees.[4]

On one occasion (December 6, 1492) Columbus says, "Sueste y tomava de la quarta del leste," literally "SE and taken from a point to the E," which means SE, Easterly.

No very accurate maps of northern Haiti exist. The best are the charts of the coast and adjacent waters issued by the United States Hydrographic Office, and are followed, at a considerable interval of time, by the British Admiralty charts. Cooke's *Carte de la Republique d'Haiti*, though useful for the interior, is unreliable, in several instances placing the mouths of rivers at a considerable distance from the points where they actually empty into the sea. Of the earlier maps of Haiti, the best is in the atlas published at Paris in 1791 by Moreau de Saint-Méry. This author's *Description . . . de la Partie Française de l'Isle de Saint-Domingue* (2 vols., Philadelphia, 1797–98) is the classic description of Haiti and often of great assistance in identifying localities today.[5]

---

[3] Dr. Willem Van Bemmelen, "Die Abweichung der Magnetnadel; Beobachtungen, Säcular-Variation, Wert- und Isogonensysteme bis zur Mitte des XVIII^ten Jahrhunderts," Royal Magnetical and Meteorological Observatory at Batavia, *Observations*, XXI (Batavia, 1899), Supplement, p. 5.

[4] Points in brackets are those that Columbus did not happen to mention during his First Voyage.

[5] This work was published at Philadelphia because the author became a refugee from Haiti during the French Revolution, and opened a book shop in Philadelphia. He was elected to the American Philosophical Society and his *Voyage aux États-Unis*, published in 1913, gives an amusing picture of social strata

L'ÉQUIPAGE.

Above: "Chaloupe No. 5," Col. Cham, and Lieut. Élie, Tortuga in background.
Below: The Crew.

Columbus's Compass Points	English Compass Points	Abbreviations	Expressed in Degrees
Norte..........................	North	N	0
Norte cuarta del Nordeste..........	North by East	N by E	11¼
Nornordeste......................	North Northeast	NNE	22½
Nordeste cuarta del Norte..........	Northeast by North	NE by N	33¾
Nordeste........................	Northeast	NE	45
Nordeste cuarta del Leste..........	Northeast by East	NE by E	56¼
Lesnordeste......................	East Northeast	ENE	67½
Leste cuarta del Nordeste..........	East by North	E by N	78¾
Leste *or* Levante..................	East	E	90
Leste cuarta del Sueste............	East by South	E by S	101¼
Lestesueste *or* Lesueste............	East Southeast	ESE	112½
Sueste cuarta del Leste............	Southeast by East	SE by E	123¾
Sueste..........................	Southeast	SE	135
[Sueste cuarta del Sur]............	Southeast by South	SE by S	146¼
Sursueste.......................	South Southeast	SSE	157½
Sur cuarta del Sueste..............	South by East	S by E	168¾
Sur.............................	South	S	180
Sur cuarta del Sudueste............	South by West	S by W	191¼
Sursudueste *or* Sursudoeste.........	South Southwest	SSW	202½
Sudueste cuarta del Sur............	Southwest by South	SW by S	213¾
Sudueste........................	Southwest	SW	225
Sudueste cuarta del Oueste.........	Southwest by West	SW by W	236¼
Ouesudueste *or* Oueste Suedueste......	West Southwest	WSW	247½
[Oueste cuarta del Suedueste].......	West by South	W by S	258¾
Oueste [1]........................	West	W	270
Oueste cuarta del Norueste.........	West by North	W by N	281¼
Ouestnorueste *or* Ouesnoroeste.......	West Northwest	WNW	292½
Norueste cuarta del Oueste.........	Northwest by West	NW by W	303¾
Norueste........................	Northwest	NW	315
[Norueste cuarta del Norte]........	Northwest by North	NW by N	326¼
Nornorueste......................	North Northwest	NNW	337½
[Norte cuarta del Norueste]........	North by West	N by W	348¾
Norte..........................	North	N	360

[1] Columbus also spells the word for West and its compounds, *Gueste* and *Vueste*.

2. CAPE ST. NICOLAS TO JEAN RABEL POINT

At sunrise December 5, 1492, the Admiral was about 7½ miles west of Cape Maisi, the eastern promontory of Cuba. Shortly after passing it he sighted Cape St. Nicolas, bearing SE; the day must have been very clear, for this western point of Haiti is distant about 44 miles from Cape Maisi. He was aboard *Santa Maria*, of which Juan de la Cosa was master, and was accompanied by the caravel *Niña*, commanded by Vicente Yañés Pinzón master. *Pinta*, commanded by Martin Alonso Pinzón, had left the fleet on November 21, and visited Babeque (Great Inagua Island) in the hope of being the first to find gold.

below that of the "Republican Court." An atlas of plates and maps to accompany the *Description* was published at Paris in 1791, and a new edition of the *Description* was printed there in 1875, with atlas. My quotations are from the Philadelphia edition.

The tradewind conveniently shifted from NE to N as the two vessels steered across the Windward Passage. They made very good speed; Columbus reckoned 6 knots until one in the afternoon, and thereafter 7½, probably a slight overestimate. The Admiral shaped his course to the SE by E because he had noticed that "always the wind veered from N to the NE, and thence to the E and SE"[6]—a common phenomenon in this tradewind belt. By sunset the vessels were a few miles off Cape St. Nicolas. The Admiral then ordered *Niña*, "because she was swifter" to precede the flagship in search of a harbor. She, arriving after nightfall "at the mouth of a harbor which was like the Bay of Cadiz," sent a boat ahead to take soundings, and to show a light to guide her in; but before the flag ship could beat up to *Niña* the boat's light went out. *Niña* then ran down to *Santa Maria* to tell the Admiral what had happened. The boat managed to light up again, and *Niña* went in after her; but Columbus, somewhat disturbed by the Indians' signal fires that appeared on distant hills, decided to wait until daylight before entering. He steered offshore about N by E with the night land breeze, in order to be able to coast down-wind to the harbor next morning.

At daylight on Thursday, December 6, the Admiral took bearings on several prominent points of land. To the S by W bore a "fine cape" which was the farthest point of Haiti to be seen to the South, and which he named *Cabo de la Estrella* (Cape of the Star). Although it is clear that he later identified this name with Cape St. Nicolas, it is probable that what he then took the bearing on was Cape Foux, about three nautical miles to the SW of Cape St. Nicolas, since it is the western tangent of Haiti. Another mass of land "like a not very big island" which he named *Tortuga* (Turtle), bore E, distant 30 nautical miles. Another "very fine and well formed" cape which he named *Cabo del Elefante*, appeared 40 nautical miles to the E by S. This cape, from what he says about it the next day, was certainly the mountain *Haut Piton*,[7] the summit of which is in lat. 19° 52′ N, long. 72° 45′ W, together with its northerly slope which ends in Grande Pointe.[8] The bearing was taken on the lowest part of the point that he could see; this was the farthest that he could then sight along the north coast of Hispaniola. Another point of land, which he named *Cabo de Cinquin* and which from his indications the next day was certainly Jean Rabel Point, lay 21 nautical miles to the ESE. And "a great fissure or opening or gorge to the sea, which seemed to be a river" bore SE E'ly 15 miles.[9] This was an illusion; there is no such opening in this ironbound coast,[10] which the French called *Les côtes de fer*. Columbus added that at the time these bearings were taken he was 4 leagues from the

---

[6] *Journal*, December 5, 1492. Assuming that they passed Cape Maisi at 9 a.m. this would mean 54 miles by 5 p.m., but it is only 44 miles straight across and the SE by E course first sailed would not have accounted for the difference. The Admiral thought they had sailed 66 miles between sunrise and sunset. Ferdinand Columbus in his *Historie della Vita e dei Fatti di Cristoforo Colombo*, ch. 30 (Caddeo ed., 1930, I, 194), more correctly says that the distance was 16 leagues (48 nautical miles).

[7] Also called *Mont Beaubrun* (Cooke).

[8] Also called Pointe du Carénage (Moreau de Saint-Méry) and Carenero (U. S. Hydrographic Office charts); but the Carénage is really inside the point.

[9] Following this passage is one that is obviously corrupted: "It appeared that between the (*entre el*) Cabo del Elefante and that of Cinquin was a very great opening, and some of the seamen said it was the separation from (*apartimiento de*) the island, to which they gave the name Tortuga." For *entre el* read *al norte del*.

[10] That is, at this bearing, or between Jean Rabel and the Mole St. Nicolas. So I am informed by Mr. Irving Crosby, an engineer who has prospected the whole coast for water.

harbor which he first called *Puerto Maria*, and then changed the name to *Puerto de San Nicolas* in honor of the Saint whose day it was.

These bearings taken by the Admiral were remarkably accurate. Four of them converge at lat. 20° 03′ N, long. 73° 24′ W, a point just about 12 nautical miles north of the harbor; and the fourth bearing, on Tortuga, also meets at the same spot if we take it to have been on the summit of the island rather than the West Point.[11] His estimated distances were pretty inaccurate, for he had nothing to measure by; no known heights of mountains, or houses, or anything else.[12]

There seems to have been a very light wind that day (December 6) for the *Santa Maria* did not close with *Puerto de San Nicolas* (now Port St. Nicolas or Baie du Môle) until "the hour of vespers" (3 or 4 p.m.). On arrival the Admiral "marvelled at its beauty and excellence," and wrote a good description, except for the dimensions.[13] "The depth of this harbor is marvellous," he rightly says. At a certain distance from the shore his sounding lead found no bottom at 40 fathoms,[14] and within that distance at a *pasada* from shore there was 15 fathoms and clear bottom; not a single rock anywhere, and so much room that "a thousand carracks could beat" to windward—not at the same time, surely! On the south side of the harbor he noted a beach, across the middle of which emptied a river.[15] Off it the *Santa Maria* anchored for the night of December 6–7, 1492.

At midnight January 14, 1939, when Orion was dipping to the West, the Great Bear was standing up on his tail, and the guards of Polaris marked the hour, our *timonier* of Chaloupe No. 5 chose a spot just off this beach, beyond the village, where we lay quietly in two fathoms while the surf roared on the Môle outside. And just before dawn I had the pleasure of seeing the Southern Cross over the hills. Our cook prepared breakfast in an open deck galley, of the same model as Columbus shipped on the *Santa Maria*; but we had some delicious Haitian coffee, a pleasure which the Admiral never experienced.

Columbus accurately describes in his Journal the inner harbor of Port St. Nicolas, which the French called Le Carénage, since it proved an ideal place to careen a ship. It had a depth of 11 fathoms throughout, he says—really 5 to 7 fathoms with a few soundings of 11—; and a ship could moor close enough to the shore to lay a gangplank onto the grass, which is true enough. He thought that the isthmus at the head of the Carénage might be cut through to form a canal. But the Admiral's high hopes for the future of Port St. Nicolas were not fulfilled. The trees which he believed to be nutmegs and "spiceries"

[11] As an example of how Columbus's *Journal* has been mangled by translators, see comparison of those bearings given in the Spanish text in several translations, in *Hisp. Am. Hist. Rev.*, XIX (1939), 235–61.

[12] For instance, Columbus said that this spot whence he took the bearings was just 4 leagues from the port, and it was just about that, as he doubtless ascertained by sailing along it before he wrote up his log for the day. But Cape Foux, which he said was 21 nautical miles distant, was really 16½; Tortuga, 30 miles, was really 37 to the summit; Cape Carnero, 40, was really 33½; C. Jean Rabel, 21, was really 14½. He estimated (in the letter to Santangel) the distance from Cuba to Cape St. Nicolas Mole as 54 nautical miles, and it was really 44; but let us not be too critical of him, because this faulty estimate is repeated on maps as late as Von Keulen's of 1734, and Faden's of 1795!

[13] Columbus says it is 1½ leagues wide at the entrance and extends SSE for 2 leagues. "Leagues" are probably errors of the transcriber for "miles," for if we substitute Columbus's Roman miles for "leagues," the figures will be nearly correct.

[14] Columbus leaves the distance blank; actually the 40-fathom line is between 300 and 700 yards from the shore up to the village.

[15] The river now disappears under the sand some 3 kilometers away from the beach.

ESTRELLA AND CINQUIN.
Above: Cape St. Nicolas seen across the Isthmus of Cape St. Nicolas Môle.
Below: Pointe Jean Rabel.

were only bayahonde (a kind of mesquite), and the country behind it is so arid and barren that even in the heyday of the French colonial régime, the harbor was important only for its fortifications.   Le Môle is now a poor, isolated fishermen's village, accessible from the land only by footpaths, and seldom visited by vessels.

On Friday, December 7, 1492, "at the relieving of the dawn watch," [16] the Admiral got under way; we followed him by half an hour on January 15, 1939.   He was fortunate, considering the season, to have a SW wind, which whipped him around Cape St. Nicolas Môle in short order, and sent him romping along the coast to the eastward.   We encountered the usual NE tradewind, but had a good gasoline motor to shove us into it; the Admiral had the advantage of us in staying dry.   He noted that the Cape was so steep on its seaward side that "a lombard's shot" [17] off shore you could find no bottom; our chart confirmed this fact.   One can appreciate when seeing the Cape from the northward why the French called it Le Môle; it is of uniform height, flat-topped, and falls off vertically like a gigantic quay.[18]   *Cabo de la Estrella*, meaning in this case the mountain behind Cape St. Nicolas, was visible over the isthmus,[19] both to Columbus and to us.   The coast from the Môle eastward is a high series of shelves or prehistoric beaches, desolate and barren of all but bayahonde and such thrifty trees, until one reaches Jean Rabel Anchorage. Columbus says that two leagues from Cabo Cinquin he saw an *agrazuela*, "a craggy spot which seemed to be an opening in a mountain, through which he discovered a very great valley," planted with some cereal crop.[20]   Perhaps this was the same "fissure or opening or gorge" which he saw bearing southeasterly at dawn, December 6.   We, possibly because we hugged the shore, saw no such opening at the distance he describes; but at Jean Rabel Anchorage a valley opens up, with banana plantations.   A Standard Fruit steamer was just departing as we approached.

Cabo Cinquin, from the bearings and courses given by the Admiral, cannot be other than Jean Rabel Point.   But one of Columbus's statements is puzzling.   "When he reached the Cabo de Cinquin," says Las Casas, "at a lombard's shot distance is a rock in the sea, which sticks up high and can easily be seen. . . . the *Cabo del Elefante* bore E by S," and the Cape of Tortuga bore NE.

Both bearings are correct, but we found no rock visible or awash off Jean Rabel Point. Columbus could hardly have invented it, and an examination of two eighteenth-century maps shows crosses—the symbol for rocks awash—just off Jean Rabel Point.[21]   Our *timonier*, who is familiar with these waters, said that there were rocks on the bottom, at least 2

---

[16] *Al rendir del quarto del alva;* presumably at 7 a.m.   Columbus usually changed his watches at 3, 7, and 11.

[17] Columbus frequently uses this measure of length.   By comparing it with known points, I have reached the tentative conclusion that it is between 800 and 1000 yards.

[18] There is a good engraving of it on Faden's Map of 1795.

[19] The point of the Cape is concealed, I observed, by the eastern face of the Môle.

[20] The *Journal* says it is "2 leagues" from the Cape; probably a scribe's error for 2 miles.

[21] *Carte réduite de L'Isle de Saint-Domingue et de ses Débouquements pour servir aux Vaisseaux du Roy . . . 1750.   Par Mr. Bellin ingénieur ordinaire de la Marine . . . Corrigée en 1764 sur les observations . . . faites en 1753.*   This shows three rocks.   *A Chart of the Windward Passage . . . with the other Passages to the Northward of Hispaniola: from the Journals, Observations and Draughts of Mr. Charles Roberts, Master in the Royal Navy, compared with the Pilote de L'Isle de St. Domingue of M: de Chastenet Puységur, 1787, and with the Descriptions des Debouquements of M: Bellin, 1768* (London, printed for Wm. Faden, 1795).   This shows one rock.

VALLE DEL PARAISO.

Above: The point reached by Columbus's boats on Trois Rivières.
Below: The mouth of Trois Rivières, Tortuga in background.

fathoms under water; but the waters off Jean Rabel are extremely rough, and I had no stomach to cruise about there and sound in order to verify his statement. The obvious explanation is that the isolated rock observed by Columbus, being of coral formation, was eaten through by the waves and fell into the sea some time during the last century.

As we rounded Jean Rabel Point, a number of capes made out to the eastward: the western points of Port à l'Écu and of Moustique Bay, and Grande Pointe, the Admiral's *Cabo del Elefante*. It did not require much imagination to see the resemblance of this last cape, with the high mountain behind it, to a huge elephant coming down to the Tortuga Channel to drink. But I rather think that the Admiral's name for it meant something more. He was doubtless familiar through the *Imago Mundi* with that passage in Aristotle's *De Caelo* where the Stagyrite remarks that the presence of elephants both in Africa and in India proves a land continuity between those regions.[22] Columbus supposed that he was in the Indies, and doubtless he expected any moment to see elephants sporting among the woods of Hispaniola.

3. MOUSTIQUE BAY, TORTUGA, AND TROIS RIVIÈRES

Columbus's Journal continues: "At 6 leagues after we found a great promontory,[23] and saw in the land behind many great valleys and plains and very high mountains, all resembling Castile." This fits the point with sand beaches just before reaching Port à L'Écu, where the back country opens up; but for leagues one must read miles. Again making the same allowance for a mistake in transcription, "Thence at 8 [Roman] miles he found a very deep but narrow river, although a carrack could easily enter it, and the mouth all clear without bar or rocks." This was Port à L'Écu, which the Spaniards named *Puerto Escudo*, evidently because its shape is that of a Spanish armorial shield. "Thence at 16 Roman miles," the exact distance from Jean Rabel, "we found a very wide and deep harbor, found no bottom at the entrance nor less than 15 fathoms at the edges three paces from the shore, and runs inland a quarter of a league." [24] This was Moustique Bay, which Columbus named *Puerto de la Concepcion* because it was the vigil of the feast of the Conception of the Virgin. Although it was only one o'clock in the afternoon he decided to anchor in this pretty harbor lying at the mouth of a valley.

The *Santa Maria* and *Niña* spent several days in this *Puerto de la Concepcion* (Moustique Bay), although it did not afford good shelter from the N to NE winds and cold, and the rainy weather that they experienced. The fleet dragged anchors on the night of December 9–10, at which "the Admiral was surprised." He had not the advantage of reading the British *West Indies Pilot*, which would have warned him "the bottom is rocky." It was here that, according to Las Casas, "seeing the grandeur and beauty of this island and its resemblance to the country of Spain although much superior and because they had caught fish like the fish of Castile and for other similar reasons, the Admiral decided on Sunday, December 9, to name this island *Isla Española*" (Spanish Isle), "by which

---

[22] *De Caelo*, ii, 14,298; W. D. Ross ed., II, 298 a; cf. Alex. von Humboldt, *Examen Critique de l'Histoire de la Géographie de Nouveau Continent* (Paris, 1836–39), I, 125–26.

[23] This word may mean a point, a bay, or an angle on the coast.

[24] The 100 fathom line comes within a straight line between the two points of the harbor, and there is a 15 fathom sounding on the chart only 70 yards from the eastern shore. Moreau de Saint-Méry (*Description*, II, 17) identifies *Puerto de la Concepcion* with Port à l'Écu; but Navarrete and all later writers have correctly identified it with Moustique Bay.

it is still known." [25] And on December 12 he caused a great wooden cross to be planted on a conspicuous rounded point on the west side of the harbor, and took formal possession *por Los Reyes y por Nuestro Señor*. How appropriate it would be to erect a stone cross there today!

It was also at Moustique Bay that Columbus made his first contact with the Indians of Haiti. They fled at his approach, but his men captured a "very young and beautiful" woman, clad only in a gold nose ring, and took her aboard the flagship, where she chatted with the female captives from Cuba. The Admiral "had her clothed and gave her glass beads and hawks' bells and brass rings" (first doubtless abstracting the gold nose ring), and sent her ashore "very honorably, according to his custom." The seamen who set her ashore reported that she would have preferred to join the Cuban ladies on their somewhat forced cruise. After this hopeful introduction the Admiral sent a squad of nine men-at-arms with Indian interpreters to make a better contact. They marched inland 4½ leagues to a village of a thousand houses [26] where the woman belonged; they made friends with the people, who presented them with bread made of yams, which they said tasted of chestnuts, and with a flock of parrots, which doubtless began their education in nautical Castilian forthwith. They heard the singing of "nightingales," the *Mimus polyglottus dominicus* or Hispaniola Mockingbird that is one of the commonest songsters of the island,[27] but were disappointed at finding no trace of gold.

During their absence the Admiral attempted an observation of Polaris with his quadrant, and worked out that he was in north latitude 34°, the parallel of Wilmington, N. C. This was a slight improvement over his earlier observations, which had placed the coast of Cuba in the latitude of Cape Cod; but still very bad.[28] The correct latitude is 19° 55'.

On Friday, December 14, the two ships "left that Puerto de la Concepcion with the land breeze, and a little after it fell calm, and thus he experienced it every day of those that he was there. Later the wind came east, he sailed with it NNE, arrived at the island *Tortuga*, saw a point of it that he called *Punta Pierna* [leg of mutton] which bore ENE of the head of the island and was distant 12 Roman miles; and thence discovered another point that he called the *Punta Lanzada* [lanced] on the same NE rhumb which was distant 16 Roman miles; and thus from the head of Tortuga up to the *Punta Aguda* [sharp point][29] is 44 Roman miles, which are 11 leagues to the ENE."

Tortuga stretched along the northern horizon as we steamed along the coast, its high, even profile reminding me of Grand Manan as seen from the coast of Maine. But the channel was rough and the gasoline getting low, so I had to forego the pleasure of visiting Tortuga. The capes are so slightly salient that they cannot be seen, much less identified,

[25] *Historia de las Indias*, lib. i, ch. 52; 1927 ed., I, 255. The native name was Haiti. The name *Bohio* also appears in Columbus's *Journal*. This apparently meant "home" in the Arawak language.

[26] This village was probably on Trois Rivières, for Columbus says that it lay to the SE of Moustique Bay and that a wide, large river flowed down the valley.

[27] Alex. Wetmore and B. H. Swales, *The Birds of Haiti and the Dominican Republic* (Washington, 1931), p. 331.

[28] The writer expects to publish shortly in *The American Neptune* an article on Columbus's navigation in which an explanation is offered for these errors.

[29] Apparently this is the same as *Punta Lanzada*. It is typical of the wild work done hitherto in tracing Columbus's route that in the map of it by Sir Clements Markham, reprinted in Filson Young, *Columbus* (3d ed., 1912), p. 236, these two points which the Admiral expressly says are in Tortuga, are made the two points of Cape Haitien!

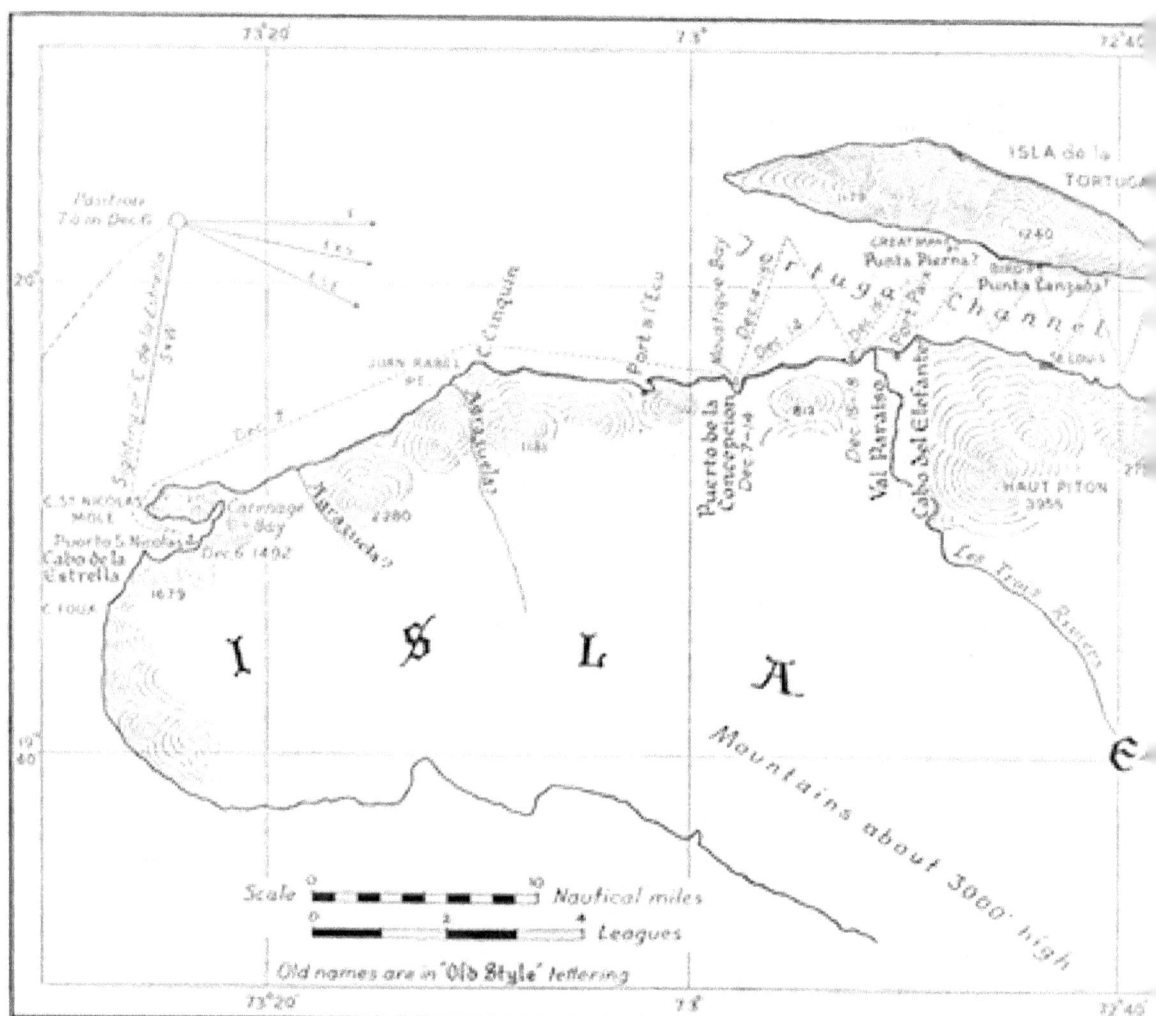

73°20' 73° 72°40'

ISLA de la TORTUGA

Position 7.15 on Deck

Position 7.15 on Deck

20°

JUAN RABEL P.te

Santiago, C. de la Galera, 548

C. Cinquin

Port a l'Ecu

Mariapue Bay

Tortuga Channel

GREAT BAY Punta Pierna?

Punta Lanzada?

SALINES

Puerto de la Concepcion Dec 7-14

Val Paraiso

Cabo del Elefante

Les Trois Rivieres

HAUT PITON 3955

C. S.t NICOLAS MOLE

Carenage Bay

Puerto S. Nicolas Dec 6 1492

Cabo de la Estrella

C. FOUX 1679

Maraticula?

2280

Maraticula 1086

I   S   L   A   E

Mountains about 3000' high

Scale 0 ————— 10 ————— Nautical miles

0 ——— 2 ——— 4 ——— Leagues

Old names are in "Old Style" lettering

73°20' 73° 72°40'

Monte Cristi Bank

THE SEVEN
BROTHERS

MONTE
CRISTI
Cabo...

R. Yaque

YUNA PT.

PT.

...RTI DU BISLUF PT.
CHOGOCHOU HEAD

MARKGOT HEAD

LIMBE PT. C. Rita y Bajo
LIMBE I. St Tomab

Dec. 24

PICOLET PT. Punta Santa

Dec. 25

Jan. 4 1493 "Niño"

BARRIER REEF

MANZANILLO PT.

BAY
La Amiga

CAPE
HAITIEN

Ponta
Caribata

Puerto
Navidad

Caracol
Bay

LES
MOWELLES

Dos
Hermanos
Tanduguda
Cove
7575

Dec. 26
Jan. 3

Vaquesi
Caracol

Puerto &. Wawas
Dec 20-25

Grande R.

Tana R.

Fort
Liberté

1442

M. Canoa R.

P   A   Ñ   O   L   A

CITADEL
2886

19°
40'

...AS TO MONTE CRISTI. By Erwin Raisz.   Copyright S. E. Morison.

except by approaching near the shore and sighting along it, as the Admiral did.   His general description of Tortuga as "very high land but not mountainous" is accurate: its crest is at a fairly uniform elevation of about 1200 feet, and presents an even skyline to the eye, unlike Hispaniola.   Apparently the Admiral's compass was not behaving well that day, since he says that the coast of Tortuga trends ENE; but curiously enough he shows the correct trend, E by S, on his chart.   Perhaps the explanation is that he sailed around the west end of the island, whence the coast does trend ENE, and that *Punta Lanzada* was its northernmost cape.   I am rather inclined to this belief, because Tortuga is far more accurately drawn on Columbus's own map than on any other before the eighteenth century.

From Tortuga, Columbus intended to visit *Baneque* or *Babeque*, where the Cuban Indians repeatedly told him the gold came from, and whither Martin Alonso Pinzón had already taken the *Pinta* on the same quest.   Baneque was either Great Inagua Island, which is not a source of gold, or a figment of the Indians' imagination.   "Seeing that the wind was contrary and they could not go to the island of Baneque, he decided to turn back to Puerto de la Concepcion."   If the wind still held to the eastward, they could have made the Great Inagua perfectly well; but of course Columbus had no exact idea of its position.   On his way back to Moustique Bay, he tried to fetch "a river which is on the E side two leagues from the said harbor"—Trois Rivières—but was unable to make it. Probably the wind flattened and the current that runs almost constantly through Tortuga Channel set him to the westward, so he returned to Moustique Bay for another night's rest.

On Saturday, December 15, the Admiral "departed from Puerto de la Concepcion once more," but on leaving the harbor was met by a contrary East wind.   He sailed on the starboard tack toward Tortuga, and then came about in order to investigate Trois Rivières, the mouth of which he had been unable to fetch the previous afternoon; "nor on this tack could he fetch it, but anchored half a league to leeward off a beach, good clean holding ground."   This place is undoubtedly the cove about one and a half nautical miles west of the mouth of Trois Rivières.[30]

As soon as the two ships were anchored the Admiral went with their boats to the river's mouth to fill water-casks.   Intending to visit the big village that the men had seen on Thursday, he had the seamen tow the boats up-stream with a hawser, but the current was so strong that they gave out after pulling "two lombard shots."   On January 15, 1939, we drove from Port-de-Paix to the very bend where the Admiral's boats were stalled, and walked down the river bank to the sea.   Columbus "saw some houses and the great valley where the villages were, and said that nothing more beautiful he had ever seen, than the river flowing through the midst of that valley."   And again, "It was a wonderful thing to see those valleys and rivers and sweet waters, and lands [fit] for crops and cattle of all sorts (of which they have none) for orchards, and for everything in the world that man can want." [31]   A beautiful valley indeed is that of the Trois Rivières, opening up the country for miles inland, with tree-clad mountains and banana groves composing a perfect

---

[30] The point between it and Trois Rivières is called Whale Point on the U. S. chart, and Pointe des Trois Rivières on the Cooke map.   The U. S. chart shows two mouths to Trois Rivières, but there is actually only one, as Columbus says.   This river rises far up in the mountains beyond Plaisance, and is joined by two other mountain streams.

[31] Journal for December 16.

setting for the stream. Columbus named it *Valle del Paraiso*, and the river *Guadalquivir*, because it reminded him of the Guadalquivir at Cordova. *Valparaiso* it remained throughout the Spanish rule in Hispaniola.

We returned to Port-de-Paix as did Columbus to his anchorage, much pleased with our excursion; and we had the additional pleasure of being entertained by Colonel Philippe Cham of the Garde de Haïti. Port-de-Paix is one of the cleanest and prettiest little towns of Haiti. A noble royal palm serves as *arbre de la liberté* in the public square, and as landmark for vessels approaching the anchorage. Port-de-Paix was the first French capital of the island, founded by Deschamps de la Place in 1664, and the seat of the first French garrison.[32] As to what peace the town was named after, I am far from clear, since its early history was notable for fights between buccaneers, French, English, and Spaniards; Colonel Cham humorously suggested that the name refers to *la paix éternelle* that now reigns over the port. Nothing to do in the evening but go to bed; which we did, after a delightful literary *causerie* over rum punches with the Colonel and his brother officers. Their knowledge of literature would do credit to any French officers' mess, and the Colonel admitted that he was translating Hamlet into French to while away the tedium. A humble commerce in *figues-bananes* has replaced the opulent *sucreries* and *caféteries* of colonial days, and only an occasional bugle-call of the Haitian Guard recalls the martial airs of Old France.

At midnight between December 15 and 16 Columbus set sail from Trois Rivières with a light land breeze, hoping to get clear of the Tortuga Channel before the tradewind made up. But all he was able to accomplish was to stand out into the channel, pick up a solitary Indian in his canoe, and anchor off an unnamed village on the Hispaniola shore, close to the beach. Colonel Cham believes that this was at Cap Rouge, almost due south of the east end of Tortuga. I hardly think the fleet got so far to windward, and am inclined to identify the anchorage of December 16 at Port de Paix.

Shortly after Columbus had anchored, about five hundred Indians appeared on the beach with their chief,[33] a youth of some 21 years, supported by a council of elders. The Indians wore in their ears and noses grains of the finest gold, which they gave away with pleasure to the Spaniards, who must have made quite a collection. They confirmed the fact that *Baneque* was the source of gold, but directed the Admiral to the eastward; doubtless they were thinking of the Cibao. The Admiral wrote, "they are the best people in the world and the gentlest; and above all I hope in Our Lord that Your Highnesses[34] will make them all Christians, and that they will be all your subjects, for as such I hold them." Yet, in his Journal of the same day Columbus makes the first suggestion of the slavery that was destined to exterminate these kind and gentle natives:

"They bear no arms, and are all naked and of no skill in arms, and very cowardly, so that a thousand would not face three; and so they are fit to be ordered about and made to work, sow and do all else that may be wanted, and you may build towns and teach them to go clothed, and to [adopt] our customs."

[32] Moreau de Saint-Méry, *Description*, I, 694–96. Moreau supposed that Columbus had been in Port-de-Paix, but it is clear from his Journal that he anchored to the leeward of Trois Rivières, and that he never visited Port-de-Paix before December 16. The English captured and destroyed the town in 1695, after which Léogane became the French capital of Saint-Domingue (*id.*, I, 702–04).

[33] Columbus calls him *Rey*, but he must have been a subordinate chief under Guacanagari, cacique of the "kingdom" of Marien that covered the whole western end of Haiti.

[34] Ferdinand and Isabella.

Two more days, December 17 and 18, were spent by the fleet at this anchorage near Trois Rivières; the first because the wind came ENE, and the second "because there was no wind, and also because the cacique had said he would bring gold." This is the first time that Columbus caught and wrote down the Indian word *cacique* (chief), which later became a title of provincial nobility in South Carolina, and has survived in modern Spanish as the equivalent of an American Tammany chieftain. This particular cacique was not the youth seen the day before, but another and cannier chief who brought a piece of gold-leaf half as large as his hand, which he cut into small pieces and peddled out to the Spaniards. When it was all gone he promised to send for more. A diversion was produced by the arrival of a canoe from Tortuga with forty men. The cacique refused to let them land, but they managed to convey word to the Admiral that he should visit Tortuga if he wanted gold, because it was nearer than Haiti to Baneque.

On December 18, the feast of the Annunciation,[35] the fleet dressed ship and fired salutes from the lombards in honor of Our Lady. At noon, when the Admiral was dining in his cabin under the *castillo* or poop of the flagship, the younger chief came aboard with his council and a retinue of two hundred, and dined with the Admiral, who was much impressed by his good manners. There was an exchange of suitable gifts, the Castilian and Aragonese standards were displayed, and a gold coin with the effigies of the Catholic Sovereigns was exhibited. The "king" was then sent ashore in proper style, while the lombards barked out salutes. Columbus watched with interest his guest being hoisted on the shoulders of his retainers, and the principal officers of his "court" bearing in triumph the Spanish gifts before him.

4. ACUL BAY

In the night of December 18–19 the Admiral took his final departure from this anchorage—possibly that of Port-de-Paix—with the land breeze. At dawn the wind turned east, and he was unable to make port that night.[36] We too got under way at dawn from Port-de-Paix, with a light land breeze blowing, and saw the sun rise over the slope of Columbus's *Cabo del Elefante*. Here there was quite a strong adverse current, running through the narrow part of Tortuga channel. A little farther on we passed over some beautiful coral gardens where the chart notes "breakers reported" off Palmiette Point, and steered well out into the channel in order to see the coast to the eastward as Columbus saw it.

Just beyond the cape with the reddish cliffs called Cap Rouge, and beyond Anse à Foleur, is a high, prominent cape that has no name on the U. S. chart, or on any of the maps I consulted. Our people in the chaloupe called it something that sounded like "Cap Oboy," which I understand is Creole for Cap au Borgne. This is the point that

---

[35] The Annunciation was then celebrated in Spain and Portugal on December 18, in order to avoid having it fall in Lent. It appears on that date in the *Regimento do estrolabio* of c. 1509, reprinted by Dr. Bensaude as Vol. I of his *Histoire de la Science Nautique Portugaise* (2d ed., Lisbon, 1924). Columbus calls it the "day of Sancta Maria de la O" which was the popular name because the anthems began with "O." There is a church of that name at Sanlúcar de Barrameda.

[36] Navarrete identifies the desired port as *La Granja* (Fond la Grange, just E of La Borgne); but it might have been anything between there and Acul Bay.

ACUL BAY.

Above: "Cabo Alto y Bajo" and the approach to Acul Bay.
Below: "La Amiga" with "Caribata," the mountains of Cape Haitien, in background.

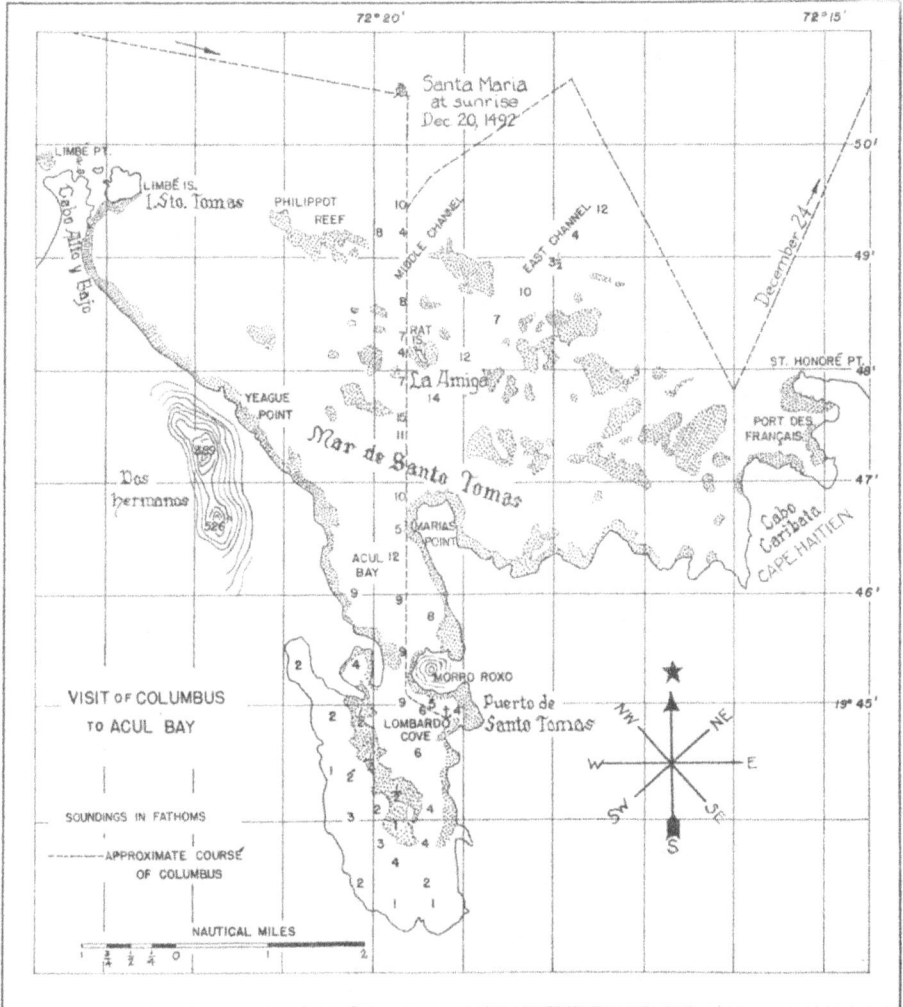

CHART ILLUSTRATING VISIT OF COLUMBUS TO ACUL BAY. By Bertram Greene.
Copyright S. E. Morison.

Columbus named *Cabo de Torres*.[37] From it, he says, bore ESE a small island which he named *Santo Tomás*, because the next day was the vigil of St. Thomas the Apostle. This bearing fits within half a point the island now called Marigot (formerly Margot) Head, and also Limbé Island; I believe he meant the former, which is the more conspicuous from that point. And from the *Cabo de Torres* there bore E by S 45 miles away "a mountain higher than the rest which juts into the sea, and which from a distance appears to be an island, owing to a *degollado* [depression] on the landward side." This bearing—but not the distance, which is 20 miles—fits the Morne du Cap, the mountain range of Cape Haitien, which, when we first sighted it from the Tortuga Channel, looked exactly like a mountainous island rising from the sea. Columbus named this mountain *"Monte Caribata*, because that province was called *Caribata*," or the Land of the Caribs, who (so he gathered from the friendly Arawaks) came from the east. Actually this region belonged to the Arawak cacique Gaucanagari, who became the Admiral's firm friend in time of need.[38]

Columbus further says that before sundown December 19 he made out "four capes of the land and a large bay and river." The four most prominent points, as we proceeded in the same direction, were Palmiste Point, East Point, Chouchou Head, and Limbé Point; and the bay may have been any of the harbors between them. "Over los 'Dos Hermanos,'" continues the Admiral, "there was a very high and big mountain which went from NE to SW." These "Two Brothers" I believe to be the mamelons of 389 and 526 feet altitude, noted but not named on the modern chart. From just west of Cap au Borgne they appear to be of equal height; and the mountain range behind them does seem to run at right angles to one's course when making for Acul Bay. Columbus also noted a cape partly high and partly low which he called *cabo Alto y Baxo*; this description exactly fits Pointe Limbé as we saw it from about longitude 72° 30′. I take it that that was about as far as the fleet had gone by sunset on December 19. Although the night of new moon, the fleet experienced no embarrassment in continuing offshore on an easterly course with the land breeze.

At sunset December 20 they entered "a harbor which was between the Island of St. Thomas and the Cape of Caribata [Cape Haitien]." This was Acul Bay, which Columbus named *la mar de Sancto Thomé*.[39] The harbor inside, he well says, is the most beautiful he had ever seen: "I have followed the sea for 23 years . . . and I have seen all the East and West, . . . but in all those regions is not found the perfection of the harbors [of Hispaniola]; . . . this [place] is superior to all, and would hold all the ships of the world, and shut in, so that one could moor with the ship's most ancient cable." [40]

Although Columbus is given to superlatives, Acul Bay is indeed one of the world's loveliest harbors, and offers such perfect security as to wring admiration from the most

[37] *Journal*, December 19. Undoubtedly named after Luis de Torres, a converted Jew, who was left with the garrison at Navidad. Another member of the family, Don Antonio, commanded the flagship on Columbus's second voyage, and another, Doña Juana, was the lady to whom Columbus addressed his famous letter in 1500. Navarrete identifies the cape as Pointe Limbé, but the ship could not possibly have sailed so far against wind and current.

[38] Filson Young, however (*Columbus*, 3d ed., p. 144), says that the Admiral anchored "for a day in the Bay of Acul, which he called *Cabo de Caribata*!"

[39] *Journal* for December 24. The inside harbor he called *Puerto de la mar de Santo Tomás*. The name Acul, says Moreau de Saint-Méry (*Description*, I, 633), is a contraction of the later Spanish name *Acon de Luysa* (Port of Louisa), a lady who lived there and was well known to sailors.

[40] *Journal* for December 21.

hard-boiled mariner.   The mountains behind compose like a landscape of Claude Lorrain, about the Bonnet de l'Evêque with King Henri Christophe's famous citadel on its summit. Lombardo Cove, undoubtedly the place of Columbus's anchorage, is completely landlocked and protected from all but westerly winds by high mountains.   It would not have held all the ships in the world, even in 1492; but the Admiral's statement about the need for nothing but old and rotten cables is not exaggerated.   In 1934 Mr. Henry Howard put in at Lombardo Cove in his yacht *Alice*, and rode out a heavy norther without even straightening out his anchor chain.[41]

Columbus was so impressed by Acul Bay that he left precise directions for entering it either from the westward or the eastward, which may be said to be the earliest sailing directions for any part of the New World.

"Whoever would enter *La Mar de Santo Tomé* should stand a good league over beyond the mouth of the entrance, over a small flat island which lies near the middle and which he named *La Amiga*, bringing the bow to bear upon it; and when one has come within a stone's throw of it, he must leave the western side [of the channel] and seek the eastern, and favor that and not the other side, since a very great reef makes out from the west (and even in the sea outside there are three single shoals), and this reef comes to within a lombard shot of La Amiga, and passing between them one will find at least seven fathoms and gravelly bottom. . . .   Another reef and shoals make out from the east side of the said island Amiga, and are very big and stretch far out into the sea, and come within two leagues of the Cape [Haitien]; but between them it seems that there is a passage, at two lombard's shots distance from La Amiga. . . ." [42]

It is interesting to compare this with the United States *Sailing Directions for the West Indies*: [43]

"Middle Channel is entered between Philippot Reef and the shoal forming the western side of the entrance to East Channel.   It joins the latter channel east of Rat Island.   A least depth of 7 fathoms can be carried through, but there is a $4\frac{1}{4}$ fathom shoal in the eastern part of the entrance and two detached shoals of much less water lie north of Rat Island.   A sailing vessel would require the wind well to the northward to use this channel, as it would have to lie up to a course of 135° [SE] and there is no room to work.   It is not recommended, as the only advantage is that the reefs are visible."

That, of course, was a great advantage for Columbus—and so it was for us.   His directions show perfectly well how he came in.   Daylight found him sailing or drifting eastward, well off-shore.   The NE trades set in as usual that morning.   As soon as the Admiral found himself "over" *La Amiga* (Rat Island), the island bearing south and covering the eastern capes of the inner harbor, he brought the *Santa Maria's* bow to bear upon it.   When fairly close to Rat Island he observed that the channel narrowed down by reason of large reefs on its western side, and that the best water lay close to the coral reef (now awash, then very probably a beach) which marks the western end of the island. His directions might be interpreted as advice to swing around to the eastward of Rat Island; but as he would have had to haul his wind to do this, and there is plenty of room between the island and the reef, I think that he continued the due southerly course, which carried him past another conspicuous reef on the west, and Marias Point and Morro Roxo on the East, into Lombardo Cove.

We joined Columbus's course at Rat Island only because, misinterpreting his sailing directions, I thought he had passed close to Limbé Point and thence brought his ship's

[41] *National Geographic Magazine*, LXXIII (1938), 301–06.

[42] *Journal* for December 24.

[43] 1936 edition, I, A, 289.

bow to bear on *La Amiga*. Our helmsman-pilot did not like this way of coming in, and insisted that I take the wheel and the responsibility. He was quite right. The straight course from Point Limbé to Rat Island brings you afoul of three small reefs NW of the island, and we had a jolly time dodging among them; fortunately it was high noon and they were conspicuous. Rereading Columbus's directions I could see that he must have come in by the Middle Channel, because he would not have been so foolish as to leave Limbé Point close aboard before daylight, and because he does not note the 11- or 12-point change of course that you have to make off the island if you come in the way we did. The last clause of the Admiral's directions refer to the East Channel, which is much the best entrance, but he probably had no chance to use it leaving port because the tradewind would not let him through.

Rat Island, Columbus's *La Amiga*, is a pretty islet covered with hardwood. On the SW side is a yellow sand beach, at the head of which are some fishermen's huts. It appears from a later entry in the Admiral's Journal that Vicente Yañés Pinzón of the *Niña*, probably when waiting for the slower *Santa Maria* to catch up with him, sent a boat ashore there and dug up some plants which he supposed to be rhubarb—the medicinal variety, not the rhubarb of tarts and pies. The Admiral was so pleased with this (since rhubarb was a drug highly prized in Spain) that later he sent a boat from Navidad all the way back to La Amiga in order to obtain a basketful of rhubarb for his Sovereigns.[44]

It was on December 20 that the *Santa Maria* and *Niña* came to this perfect anchorage in Lombardo Cove, Acul Bay. On the 21st the Admiral went in his ship's boats to explore the bay, came upon an Indian village where he admired the "very pretty bodies" of the women, who went "naked as their mothers bore them" except for "some little cotton things . . . like a flap of a man's drawers." He accepted presents of gold, parrots, nuts, fruits, and water in earthenward jars; and in return gave beads, hawks' bells and brass rings besides entertaining visitors in *canoas*. He reckoned that a thousand people came aboard by canoes, and five hundred more by swimming. The next day—the 22nd—he made sail at dawn, but returned to his anchorage when the wind turned foul, and received an embassy with gifts and an invitation to stay from Guacanagari, who lived on the other side of Cape Haitien. It took a good part of the day before the Admiral understood the invitation, which was delivered in the sign language. On Sunday the 23d there was no wind, so the Admiral sent three messengers in ships' boats, piloted by the cacique's state canoe, to call upon Guacanagari. They returned that same night, reporting that the course to the royal residence was clear, and that it would be a good place to keep Christmas. Unfortunately the Admiral agreed.

5. THE TRAGIC CHRISTMAS

On Monday, December 24, the Admiral got under way from Lombardo Cove, Acul Bay, before sunrise, with the land-breeze. He had discovered for himself a procedure recommended by the modern *Sailing Directions for the West Indies* to all sailing vessels on this coast. You leave harbor with the land breeze before sunrise, in order not to be headed by the tradewind drawing in from the north and east.

---

[44] *Journal*, December 30, 1492, and January 1, 1493. This was not the real Chinese or Turkish rhubarb, which does not grow in Haiti, but probably the plant called *Roioc* or *Fausse Rhubarbe* in O. P. Nicolson. *Essai sur l'Histoire Naturelle de l'Isle de Saint-Domingue* (Paris, 1776), p. 302.

Apparently Columbus left the bay by the same Middle Channel by which he had entered. We, on January 16, 1939, took the passage along shore inside the reefs, and had a good look into that old buccaneers' hole, Port des Français, which Columbus observed from a distance. It is shut off from the interior by the high mountains of Cape Haitien, Columbus's *Monte Caribata* or *Caribatán*.

The wind was light and unfavorable throughout December 24, 1492, the two ships had to make long tacks, and nightfall Christmas Eve found them well off Point Picolet, which the Admiral named *Punta Santa* because of the approaching holiday. At 11 p.m., when the flagship had made by the Admiral's estimate one league beyond this eastern point of Cape Haitien, he stretched out to get some sleep, and left the helmsman in complete charge. This was an unseamanlike and dangerous procedure, for on ships such as the *Santa Maria* the helmsman could not see where he was going, and had to be conned by someone from the deck above. At the time Columbus went below, the flagship was probably more than a league beyond Point Picolet, but off soundings; a reference to the map will show that the hundred-fathom line bends shoreward at that point, and the ship's ordinary lead-line was forty fathoms. The wind was light, and probably from the westward, since the flagship made a little less than two miles during the next (and last) hour of her life, supposing the Admiral's estimate of one league from Point Picolet at 11 p.m. to have been correct. The light wind died away so that the ship made little or no headway, and although there was a groundswell a calm slick was over the sea, no breakers were heard inside the barrier reef, and the five day old moon, having set about 11.30, revealed no suspicious ruffle where coral reefs concealed their teeth under a smooth surface. The caravel *Niña*, ahead as usual, was probably showing a light from her stern. So the helmsman himself decided to knock off, and handed the tiller to a young grummet, or ship's boy. This, says Columbus, had been strictly forbidden and never before practised on the voyage. And so it happened that at midnight, just when the non-existent lookout should have struck eight bells to usher in Christmas Day, the *Santa Maria* slid gently onto a coral reef, where she stayed.

A careful consideration of the Admiral's course and distance—one league or more eastward of Point Picolet at 11 p.m., and another hour of slow sailing in the direction of Guacanagari's settlement, together with a personal investigation of the reefs and of other factors that will be considered presently, has led me to the following conclusion. The *Santa Maria* cleared Point Picolet about a league offshore,[45] left the coral reefs east of it well to starboard and the barrier reef (which must have been breaking) fairly close to port, and followed the *Niña* in attempting to negotiate the channel between the barrier reef and the site of the present buoys. She struck on one of the first three of the buoyed reefs indicated on the modern chart just south of Limonade Pass. Probabilities are about even between the fatal reef being the middle or largest one, where there are now one or two fathoms of water, and the easternmost of the three, where there are two. *Santa Maria* drew fifteen feet, more or less. All three reefs, as well as the outside barrier reef, were breaking heavily on both days that we examined them, when a fresh tradewind was blowing; and the passage between them and the barrier reef seemed very narrow. But Columbus says in his Journal for January 4, 1493, that he left the bay on a NW course by a wider

---

[45] The Admiral later says (*Journal*, January 4) that there are only 8 fathoms a league off Point Picolet. A sounding of that depth is found at that distance on the modern chart. Hence I presume that the Admiral passed the point a league offshore on December 24.

channel than that through which he entered.   He must then have entered the bay to the northward of the buoyed reefs, the largest of which is the most likely to have tripped him.

This larger or middle reef is 5 nautical miles from a point a league directly off Cape Haitien.   The *Santa Maria* had made three miles by 11 p.m., and it would seem that she could hardly have sailed and drifted more than two miles more during her last hour; for it was a flat calm when she grounded at midnight.   On the other hand, the third reef (counting from west to east) fits in better with Columbus's statement that Guacanagari lived "about a league and a half from" the reef of the wreck.[46]   As we shall see later, the cacique's village must have been either at Caracol or at Yaquezi, with the probability in favor of the former.   Now Caracol is almost exactly two leagues from the larger or middle reef, 5 miles (1⅔ leagues) from the next reef to the eastward.[47]   The Admiral was not accurate at measuring distances, but there was so much rowing back and forth between the wreck and the Indian village that he had every opportunity to get this one right. He must have underestimated somewhere.   Either the distance sailed from Point Picolet or the distance from the reef to the village was more than he thought.

There is also the possibility that the *Santa Maria* met her doom on a favorite ships' graveyard, the westerly tip of the barrier reef that marks Limonade Pass.   But the barrier reef always breaks with the slightest swell, and as the flagship was following the *Niña*, which found good water, there was less chance of her running afoul of the barrier than of the reefs inside.   If the *Niña*, for instance, was alongside the third reef at midnight, the *Santa Maria* following her could well have been tripped up by the middle reef.   The little *Niña* could have sailed right over the middle reef without striking; but she would certainly have piled up on the barrier reef, where there is only a foot or two of water.

The flagship grounded so gently that none of the sleep-deadened crew felt the shock; but the boy who held the tiller "gave tongue, at which the Admiral jumped up," and then the master, Juan de la Cosa.   The Admiral ordered him and the men to haul in the boat, which was towed from the poop, and to warp an anchor out astern in the hope of kedging her off.   But they, instead of doing what was ordered, merely pulled away to the *Niña* which was lying "half a league to windward"—i.e., to the eastward.[48]   Vicente Yañés very correctly refused to receive them, "and therefore they returned to the ship; but a boat of the caravel reached her first.   When the Admiral saw that they were fleeing, and his own people at that, and the water growing shallower,[49] and that already the ship lay athwart the sea,[50] seeing no other remedy, he ordered the mainmast to be cut away

[46] *Journal* for December 25.

[47] The fourth buoyed reef on the chart, counting from the westward, fits in only 1⅓ leagues from Caracol and 1½ from Yaquezi, but it is 2½ miles further from Point Picolet than the larger reef.

[48] I do not take this to mean that the wind was east, but that Columbus had fallen into the habit that all seamen do in the tradewind belt of using "windward" and "eastward" synonymously.

[49] *y las aguas menguavan;* here I follow Cecil Jane's translation from Ferdinand Columbus, who gives it in direct quotation: *ahe scemavano le acque.*   Markham translates it "the water rising," which it cannot be.   I suppose it to mean that the ship was being driven higher on the reef by the swell, so that as the Admiral sounded along her sides, he found the depth diminishing.   It is true that the tide was falling. The United States Hydrographic Office has calculated for me that high water that night at Cape Haitien came between 10:30 p.m. and midnight.   But as the range of tide at that phase of the moon is only two feet, there would not have been a noticeable drop in half an hour's time.

[50] *y estava ya la nao la mar de través.*   I take this to indicate that she went on the reef bow on heading South, and that the swell swung her stern around so that she lay diagonally across the waves.   This clause is not found in Ferdinand's *Historie* (ch. 32; Caddeo ed. I, 203).

and the ship to be lightened as much as they could to see if they could get her off; but as the water continued to grow shallower,[51] they could do nothing, and she lay on her beam ends across the sea (although there was little or no sea running)[52] and then the planking[53] opened, but not the ship." The Admiral then went aboard the *Niña* in order "to place in safety the people of the ship on the caravel, and as already a light land breeze blew up, and there still remained much darkness, nor did they know how far the shoals extended, he lay to until daylight,[54] and then made tracks for the ship from behind the line of reef,"[55] first sending a boat to tell Guacanagari what had happened.

As this is one of the most notable shipwrecks in history, we may be permitted a bit of comment. The Admiral's Journal, which is the only account extant, suggests but does not directly state that the major blame attached to Juan de la Cosa, master of the ship, and owner of her as well, at least in part. As master it was his responsibility, not the Admiral's, to see that watches were set and that proper order and discipline were maintained. He had no business to turn in and leave the deck in charge of the Admiral and a helmsman. When the ship struck, he showed gross insubordination, and a total want of seamanship, if not of common courage, in going off with the only boat,[56] instead of warping an anchor out to windward. Thus he wasted the only chance of hauling her off. He may have thought it too late already, with every swell lifting her higher on the reef, and swinging the stern around; but it was his plain duty to try, even if the Admiral had not so ordered. Columbus uses the hard word *traicion* (treason) to describe Juan de la Cosa's disobedience, and ascribes it in part to the fact that the boat's crew were Galicians, like the master.[57] The discovery of gold always brings out the worst traits in human nature; and reading between the lines of Columbus's Journal, I think that he suspected that wrecking the *Santa Maria* was a put-up job on the part of the Galicians, in order to be left in Hispaniola and have first whack at the gold mines. Perhaps that is why Juan de la Cosa was not accorded the questionable privilege of remaining at Navidad.

By his own showing, the Admiral was not clear of blame. He excuses himself for turning in because he had lost two nights' sleep already, and "felt secure of reefs and rocks, because on the Sunday when he sent the boats to that King [Guacanagari] they had passed a good $3\frac{1}{2}$ leagues to the East of *Punta Santa* [Point Picolet] and the seamen had viewed the whole coast and the reefs that extend a good 3 leagues to the ESE of the said *Punta Santa* and saw where they should go, which had not happened on this voyage before." These seamen on their boat expedition would naturally have hugged the shore rather than run the channel between the reefs, and one of these seamen must have been

---

[51] See note 49.

[52] Columbus in this clause seems conscious of his inconsistency in saying that the sea was as smooth as a porringer of water, yet enough swell running to drive her on hard and fast.

[53] *conventos.* Las Casas, *Historia de las Indias*, lib. i, chap. lix (1927 ed., I, 278), says that this word means *los vagos que hay entre costillas y costillas*, "the planking which is between the ribs." Cecil Jane translates it "hatches"; Markham "timbers," which of course would have held longer than the planking. I am told that a coral reef punches holes in planking more quickly than any other kind of rock, but hope never to have an opportunity to make the comparison.

[54] Why didn't she anchor? There is no deep hole in the bay. Probably the lead line showed a coral bottom, and the Admiral feared losing ground tackle.

[55] This seems to mean that he approached the wreck by rowing around the southern side of the reef.

[56] The Admiral's *Journal* of January 2, 1493 proves the *Santa Maria* carried but one boat at this time.

[57] *Journal*, December 26.

piloting the *Niña*, which was ahead. Yet, even if the helmsman had been one of those who made the boat journey, there should have been at least one other A.B. on deck to con him, and a competent lookout in the bows, when navigating waters by night so full of reefs. No doubt the Admiral was exhausted from two nights' entertainment of hundreds of clamorous and importunate visitors at Acul Bay; but every master mariner must learn to go without sleep if necessary. Finally, the Admiral could well have stayed aboard the wreck on a calm night, instead of going aboard *Niña* with the crew. When he did that he threw up the sponge.

It must have been a hellish night for all, with shouting and confusion and every form of oath and reproach known to the Castilian tongue; and perhaps none of us would have kept our heads under those circumstances. No doubt the poor grummet at the helm came in for more than his share of blows and curses, although he was the least blameworthy of those concerned. But I cannot help thinking that, bad as La Cosa's action was in deserting the ship, the fact that he could have even thought of such a thing indicates that Columbus neither kept good discipline nor inspired a proper confidence on the part of his officers. The accident was inexcusable, and the behavior of the ship's company made it irretrievable.

At daylight the Admiral returned to his stranded flagship, and also sent a boat to apprise Guacanagari, whose village lay "about a league and a half from the said shoal." The cacique wept when he heard the news, "and sent all the people of the town with very big and many canoes to discharge everything from the ship; and so it was done and he cleared the decks in a very short time." Even that was insufficient to float *Santa Maria*. Guacanagari put guards over the cargo, and on the next day (December 26) when it was decided to give her up for lost, the Indians helped the crew to save everything— "he had not lost a lace-point, a board or a nail, because she remained whole as when she set sail, except that she was cut and razeed somewhat to get at the vessels and all the merchandize, and they brought everything ashore, and well guarded." But there was enough of the hull left on the day before Columbus intended to sail, when the *Niña* used it for target practice in order to impress the natives.

6. NAVIDAD AND THE INDIAN VILLAGE

The day after the shipwreck Columbus gave orders for the construction of *una torre y fortalyza* (a tower and fortress) which he named *La Villa de Navidad* (The Town of the Nativity), and the anchorage or harbor in front of it, *El Puerto de la Navidad*. It had been his intention to leave a garrison in Hispaniola, and now was the obvious time to do it, when he had the *Santa Maria's* crew on his hands, and only the little caravel *Niña* to sail home in.[58] The fortress of Navidad was constructed of the "boards, timbers and nails of the ship";[59] and the Admiral ordered that it include a "great cellar" to store the trade-goods, food and munitions. After the *Niña* sailed the garrison built a palisade around the main building,[60] which seems to have been a sort of blockhouse thrown together from the ship timbers. Columbus manned the fort with about forty men under three

[58] The *Pinta* was only reported the next day, and Columbus could not count on meeting her before leaving the Indies.

[59] B. de Las Casas, *Historia de las Indias*, lib. i, chap. 61; 1927 ed., I, 285.

[60] Dr. Chanca's letter on the Second Voyage, in Navarrete, *Colección de Viages*, I, 213.

COLUMBUS 1493

tortuja

jan nicolab

monti crifti

nativida

la ·ſpañola

JUAN DE LA COSA 1500 ?

punta de cuba

p. de s. vicbla

coeli

y tortuga

C. Snt

navida

o
X: m. P

vegareal

Isabela

c de estrella

MAP OF 1516

GUANEI

(Cap Haitien)

(Monte Cristi)

ISABLES

GUAHACANO

JAQ. FL.

MAUTION FL.

FORTIUS REGALIS

MARAGON

JAQUE FL.

RECENT CHART

73°

72°

Tortuga I.

20°

Juan Rabel Pt.

Port Paix

C. au Borgne

Marigot Head

Limbe Pt.

Cap Haitien

Caracol Pt.

Jaquesi Pt.

The Seven Brothers

Monte Cristi

20°

C. ST. NICOLAS
MOLE

2280

Moustique
Bay

Les Trois R.

3955

3932

20824

2723

Manzanillo Pt.

Acul Bay

Caracol
Bay

Fort Liberté
Bay

(Dajabon)

Massacre R.

Trou R.

Grande R.

Nord du Cap R.

Gonaives
Bay

THREE EARLY MAPS OF HAITI COMPARED WITH MODERN CHART.

Sketches made to same scale.

262

lieutenants,[61] and left with them all the trade goods, seeds, a year's provision of bread and wine, and the ship's boat. The garrison, evidently selected with care, included a ship's carpenter, caulker, cooper, gunner, physician, and tailor.[62] By January 2, 1493, the fort was completed and on the 4th the *Niña* sailed for Monte Cristi.

There has always been considerable uncertainty over the site of the Villa de la Navidad, as Columbus calls this first European settlement in the New World. The site of Guacanagari's village is also disputed. We shall first quote the evidence, and then try to piece it together, with reference to the modern map.

a. Evidence of Early Maps

Only two ancient maps, to my knowledge, place Navidad. The first is the Admiral's own outline sketch of the coast from Cape St. Nicolas to Monte Cristi, made in 1493, on which he enters the name in its Italian form, *Nativida*.[63] Let us compare it with a modern map of the same longitudinal scale; for the vertical scale, on most early maps, greatly exaggerates the capes, bays and other details that the map maker had visited. East of Tortuga we have three small islands representing La Amiga with her attendant reefs, and Acul Bay. Next comes a cape of which the further side trends southeasterly. The cape is obviously Cape Haitien, but is the point of it meant for Point St. Honoré,[64] and the southeasterly trending coast for the northern shore of Cape Haitien with Point Picolet flattened out? That would bring *Nativida* at Petite Anse. Or does the point represent Point Picolet, and the coast line the outer line of the reefs of Cape Haitien Harbor? That would bring *Nativida* to Limonade Bord-de-Mer. I believe that the latter is the correct explanation. For, if we read Columbus's map from East to West, starting with Monte Cristi, the next point is Manzanillo, the next bay to the westward of it is Caracol Bay, and the rather broad point westward of that bay, Caracol Point. Otherwise that part of the Admiral's map east of *Nativida* does not make sense. He drew his coast line around the Cape Haitien reefs to indicate that he knew of no safe entrance to Cape Haitien Harbor.

The World Map of Juan de la Cosa, dated 1500 but which Mr. George E. Nunn has argued very plausibly to be a 1508 copy of a 1500 map,[65] exaggerates the length of capes and the depth of bays even more than did the Admiral. Point Picolet on Cape Haitien is identified by *C S[a]nt[o]*, Cape Haitien harbor is shown, with a nick in it to indicate either the mouth of the Rivière Haut de Cap, or, as is more likely, the Grande Rivière.[66] Next to the eastward comes Point Sable, and then another and deeper nick against which is the name *Navida*. This second nick in the coast might be the mouth of the Grande Rivière, but it may quite as well represent the mouth of the Fossé, or the salt estuary at Limonade Bord-de-Mer. Off it one sees two reefs, a large and a small—

[61] The list is in Navarrete, II, 19–20, but Miss Alice B. Gould has shown in her series of articles on Columbus's crew ("Nueva lista documentada de los tripulantes de Colón en 1492," *Boletin de la Real Acad. de la Historia*, vols. LXXXIV–XCII, Madrid, 1924–28) that the list is not altogether accurate.

[62] *Journal*, December 26 and January 2.

[63] This is one of the few instances of Columbus using Italian.

[64] The northwestern point of Cape Haitien, corresponding to Point Picolet, the northeastern.

[65] *The Mappemonde of Juan de la Cosa* (Jenkintown, 1934). I have used the Karpinski photograph of the Map as the basis of this sketch.

[66] The Grande Rivière emptied west of Point Sable in colonial times, to judge from Moreau de Saint-Méry's *Description*.

Juan de la Cosa remembered them very well! Caracol Point is very much flattened out, but Caracol Bay is prominent, and shows a nick at the landing place. Next to the eastward the thumb-shaped point represents Yaquezi. Fort Liberté Bay, although discovered in December, 1493, is omitted; Manzanillo Point and m: Χρο (Monte Cristi) form the SW and NE points of a hammer-headed cape.

b. The Topography

Now for the modern topography. No accurate map of the region between Cape Haitien and Monte Cristi has been published. The coast from Petite Anse to Fort Liberté Bay has never been properly sounded or surveyed, and its coast line is very inaccurately represented on the most modern United States Hydrographic Office chart. Cooke's *Carte de la République d'Haïti* (1928), the official government map, is also inaccurate, representing certain rivers as emptying at the wrong points, and is of too small a scale to be useful; the sketch of Caracol Bay in S. Rouzier, *Dictionnaire Géographique et Administratif d'Haïti* (Paris, 1891) is even worse. Fortunately I have had access to a photographic air map of the region between Fort Liberté and Limonade made by Lieutenant Wayne Boyden of the United States Marine Corps in 1926, on a scale of 1 to 10,000 (over 6 inches to the mile); and the sisal planters who lent it to me, Mr. Robert L. Pettigrew of Fort Liberté, and Mr. Kenneth Crooks of Cape Haitien, have accompanied me on personal reconnaisances both by land and sea. The accompanying sketch map is based on the Hydrographic Office chart for reefs and soundings, and on Lieutenant Boyden's air photograph plus my personal investigations for dry land and mangrove swamp.

During the eighteenth century, Cap Français (Cape Haitien), the "Paris of the Antilles," was the principal city and seaport of the wealthiest French colony, and the center of an area of incredible fertility. The Vega Real, stretching from the sea about ten miles up to the mountains, was thickly cultivated with sugar, coffee and indigo, which was exported to France and to the English Colonies from Cape Haitien. The barrier reef, affording protection both from the sea and from enemies, enabled the planters east of Grande Rivière to convey their produce safely and cheaply to Cape Haitien by small boats, the shipping points being Limonade Bord-de-Mer, Caracol, and Yaquezi. All three places were fortified, lest a bold enemy ship force the passes in the reef, and break up the traffic. It was even proposed to link up Fort Liberté Bay with Caracol Bay by a canal, in order to bring it within this system of protected inland waterways.[67]

Then as now, an important topographical feature of the bay was the mangrove swamp that fringes the shore. By far the greater portion of the shore line from Petite Anse near Cape Haitien to Les Mamelles near the entrance to Fort Liberté, is composed of mangrove trees growing in mud flats covered by a few inches of salt water. These completely obstruct access by boats to the dry land; consequently the only possible landing places are at gaps in the mangrove. Of these gaps there are only three: the first, about three miles long, from Point Sable along a sand beach to a point a few yards east of Limonade Bord-de-Mer; the second, about half a mile long (with a belt of mangrove in the middle) at Caracol; and the third, a narrow creek through the mangrove at Yaquezi.[68] Also, the shores of Cape Haitien harbor are free of mangrove to a point just east of Petite Anse.

[67] Moreau de Saint-Méry, *Description*, I, Partie du Nord, chs. 4–6, esp. pp. 160–63.

[68] Mr. Crooks has built an artificial *embarcadère* for his plantation on Bekly Bay by cutting through the mangrove swamp to dry land.

THE SANTA MARIA REEF, AND GRANDE RIVIÈRE.

Above: Sea breaking on the Reef in Caracol Bay where the *Santa Maria* was probably wrecked.
Below: Point on Grande Rivière near which *Santa Maria* anchor was found.

The question at once arises, were the shores of this bay the same in 1492 as today? Isn't it possible that the mangroves have thrust out in the last 450 years? The entire north coast of Hispaniola has been rising for several millennia,[69] but not fast enough to account for any appreciable extension of the mangrove belt. It is true that neither Columbus nor any of his companions mention the Caracol Bay mangroves directly.[70] The Admiral, in fact, says, "all that coast runs NW–SE and is all beach [playa], and the land very flat, as far as 4 leagues inland." [71]   But on his map of 1493 the points of Caracol and Yaquezi, now solid mangrove swamp, are clearly shown. And the extent of mangrove swamp has not widened appreciably in Fort Liberté and Caracol Bays since Moreau de Saint-Méry's description and map of the 1780's.[72]   Caracol was so named by the Spaniards because of the snail-like channels in the mangrove. The fact that the sand strip behind the mangroves is at sea level, being covered in part by ordinary tide sand wholly covered by spring tides, proves that there has been no change of level, and therefore no chance for the mangroves to grow out into the bay in the last 450 years.[73]   Only at the mouth of the Grande Rivière has there been an appreciable extension of the land northward, owing to the large quantities of silt that it brings down from the mountains.

This land is indeed very flat, as Columbus noted; it is the alluvial plain of several rivers that rise in the mountains and discharge into the bay, sometimes breaking new channels and finding new mouths when in flood. The beach mentioned by Columbus is the strip of level sand without vegetation that begins where the mangroves leave off, and which varies in width from a few yards to half a mile. This strip is locally called les salines, because some of it was used for salt pans. On its shoreward side, where it is high enough to escape salt water in spring tides, the beach has grown up with bayahonde (mesquite) trees. Behind the bayahonde belt is rich, deep alluvial soil, a good part of which is still cultivated by the negroes. In colonial days this Plaine de Limonade, as the central part of it is called, was the seat of some of the richest French sugar plantations, and today one encounters everywhere impressive ruins of the massive brick and stone bridges, houses, warehouses and sugar-mills of the wealthy creoles who were wiped out when the French Revolution swept over Haiti. The parish of Limonade alone, just before the French Revolution, had thirty-seven sugar mills with an annual production of eight million pounds

[69] Charles Schuchert, Historical Geology of the Antillean-Caribbean Region, or the Lands Bordering the Gulf of Mexico and the Caribbean Sea (New York, 1935); T. W. Vaughan, et al., A Geological Reconnaissance of the Dominican Republic (Dominican Republic. Geological Survey. Memoirs, vol. I, Washington, 1921).

[70] In his Journal for February 14, 1493, however, Columbus says that he is setting adrift in a barrel an abstract of his log in order that (among other reasons) the Sovereigns might learn that there were "no tempests whatever in those regions [las Indias] (which he says you can tell from the plants and trees which are planted and growing even within the sea)." He must have seen mangroves on some of the Cuban bays, but Caracol Bay was more exposed.

[71] Journal, January 4, 1493. Mistranslated by Markham "a quarter of a league." The plain is about 12 miles wide at its widest point, and nowhere less than 8.

[72] Description, I, 161–62, and Atlas, planche 21. From his description of the coast-line of Caracol Bay, it is evident that Caracol Point extended as far north in his day as in ours, and consisted of mangrove swamp. The dimensions he gives for Limonade and Bekly Bays (I, 206) indicate that they covered the same area as today.

[73] Letter of Professor Reginald A. Daly.

of white sugar, seven rum distilleries, three indigo works, and one hundred sixty coffee plantations.[74]

The coast line on Columbus's own map, and, less decisively, those on the La Cosa and Bologna maps, indicates that the belt of mangrove swamp was roughly identical with what it is today. If that be true, the spots where Columbus could have landed and built a fort, or the Indians could have built a village, are limited. Mangrove swamps were attractive to the Arawaks by reason of the multitude of oysters that can be plucked from their roots; but these Indians were not lake dwellers, and wanted dry ground for their houses and fertile soil for their gardens. I have searched this coast with a powerful glass from the bridge of the Royal Netherlands S.S. *Calypso*, which anchored inside the easternmost buoys; I have followed it in motor boats, approached it from every possible landward point of access, and carefully studied the airplane photograph, which is of a scale sufficient to pick out the tops of individual trees. As a result, I am satisfied that there is no dry land between Cape Haitien and Les Mamelles except at the four places already mentioned, which the colonial French used as embarcadères: Petite Anse, the beach between Point Sable and Limonade Bord-de-Mer, Caracol, and Yaquezi. We may safely confine the search for the sites of Navidad and of the cacique's village to these four gaps in the mangrove.

Now for the documentary evidence.

c. Evidence from Columbus's Journal

December 23. The Admiral says that Guacanagari's village is "in the direction of" (*de parte de*) Punta Santa (Point Picolet) "about 3 leagues to the SE." This was before he had been there, and he revised the estimate two days later.

December 25. Guacanagari lived "a good 3½ leagues" East of *Punta Santa* and "about one league and a half from" the reef where the *Santa Maria* was wrecked. Caracol is about 3¾ leagues from Point Picolet, and almost exactly 2 leagues from the reef where we believe the *Santa Maria* was wrecked. The Admiral sent a boat ashore to tell the cacique that he had been shipwrecked, which suggests that the Indian village was screened from the reefs by Caracol Point. The salvaging of the cargo began, and the nearest landing place to the reef was the site of the present Limonade Bord-de-Mer.

December 28. Guacanagari observed the Admiral going ashore from the *Niña*, and the rest of that day's Journal suggests that the Indian village was not far from the spot where the Admiral had begun to direct the building of Navidad. But it is not incompatible with the village being some distance from Navidad, the visit being made by boat or canoe through one of the several channels in the mangrove; or with the cacique having observed the *Niña* from his canoe.

December 30. The Admiral remarks that the island *Amiga* is 6 leagues (19 miles) distant. It is actually 20 miles from the reef where we suppose that the *Santa Maria* was wrecked, and about the same distance from the anchorage between that reef and Limonade Bord-de-Mer.

December 31. The people were occupied in taking wood and water aboard. Fresh water could doubtless be had from the Fossé River, which (according to Moreau de Saint-Méry) was formerly a more considerable stream than in his day or today; and the season

---

[74] Moreau de Saint-Méry, *Description*, I, 195–96.

MAP TO ILLUSTRATE THE PROBLEM OF THE SITE
New World

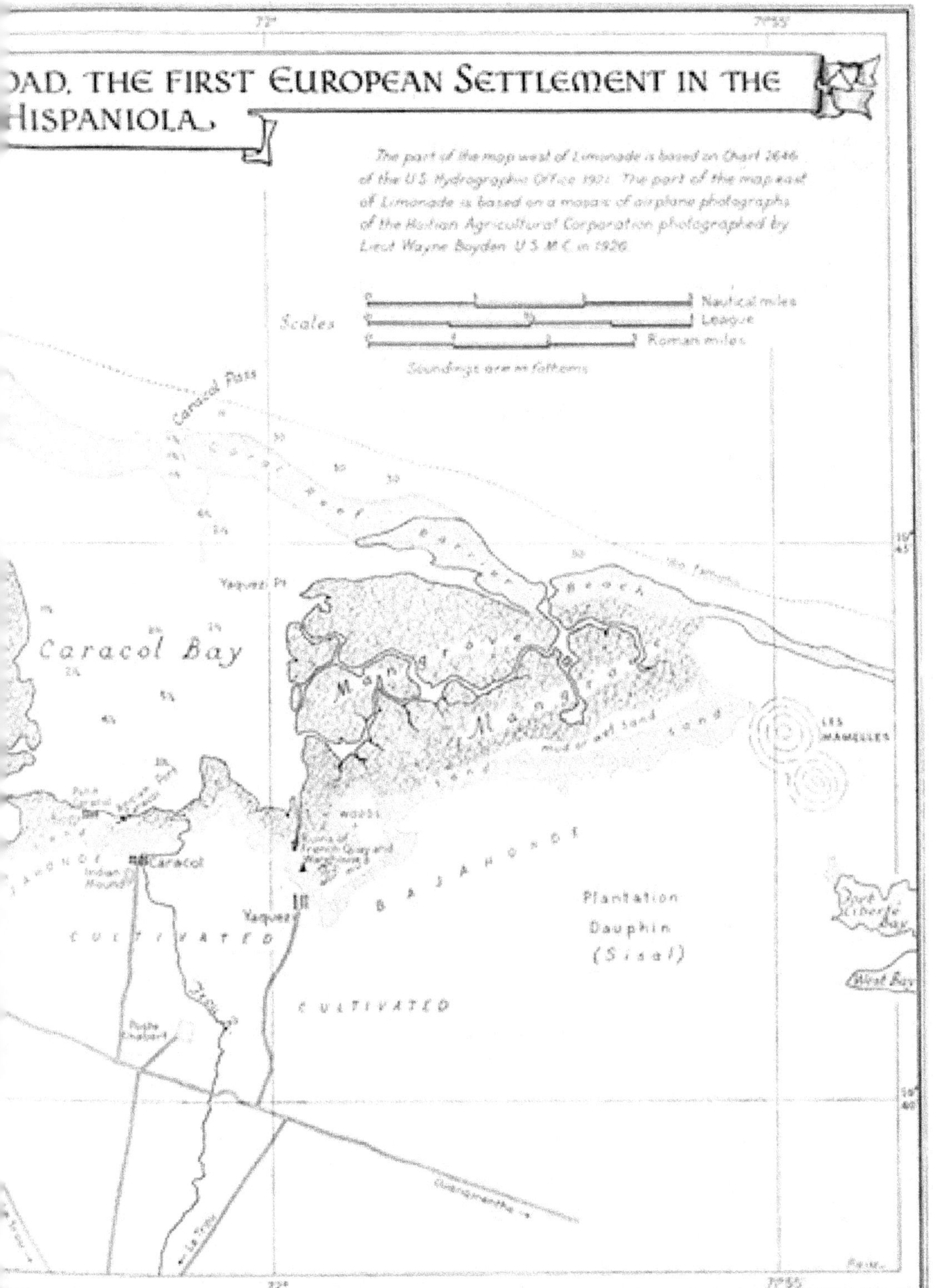
…OAD, THE FIRST EUROPEAN SETTLEMENT IN THE
…HISPANIOLA
The part of the map west of Limonade is based on Chart 2646 of the U.S. Hydrographic Office 1921. The part of the map east of Limonade is based on a mosaic of airplane photographs of the Haitian Agricultural Corporation photographed by Lieut Wayne Boyden U.S.M.C. in 1926.
Scales
Nautical miles
League
Roman miles
Soundings are in fathoms
Caracol Pass
Yaquezi Pt
Caracol Bay
Caracol
Yaquezi
Plantation
Dauphin
(Sisal)
CULTIVATED
CULTIVATED
LES MAMELLES
Port Liberté Bay
West Bay
72°
71°55'

was rainy. As for wood, there must have been plenty of good hard bayahonde growing along Limonade beach then, as now. Mangrove wood is not good for fuel.

January 2, 1493. Columbus stages an artillery exhibition for Guacanagari's benefit, the *Niña* firing shots through the remains of *Santa Maria's* hull on the reef. This again suggests that the wreck was visible from the village; but of course the cacique could have been paddled out in his state canoe to see the show.

Columbus left the ship's boat with the men at Navidad, and instructed them to search out a new place of settlement, since Navidad was "not his idea of a harbor." [75] The next day he liked it still less.

January 3. The sea was "somewhat rough," so that *Niña's* boat could not go ashore. Probably a norther had struck in. In a N to NW wind the anchorage at Limonade Bord-de-Mer has no protection, and the fishermen beach their boats in order to avoid parting their cables.

January 4. Columbus gets under way on a NW course "to get clear of the [barrier] reef by another channel wider than that by which he entered; the which and others are very good for coming in front of [*por delante de*] the town of Navidad." One may verify on our map that the only possible course fitting this description starts from near the $5\frac{1}{2}$ fathom sounding with mud bottom, which is just "in front of" Limonade Bord-de-Mer. The Admiral goes on to say that both the channel by which he entered on December 24 and the one by which he departed on January 4, "extend from NW to SE the full length of the reefs, and stretch from *Cabo Santo* [Point Picolet] to the *Cabo de Sierpe* [76] which is more than 6 leagues, and outside in the sea [they extend] a good 3 [leagues] [77] and beyond *Cabo Santo* a good 3 [leagues],[78] and beyond *Cabo Santo* for one league there is no more than 8 fathoms depth,[79] and within the said cape, on the E side, there are many shoals and channels to enter by them; and all that coast runs NE–SE."

January 8. Columbus explores the Rio Yaque del Norte, "to the SSE [i.e., SSW] of Monte Cristi a good league" (correct), and remarks that from there to the Villa de la Navidad is 17 leagues. That distance if measured in a straight line would take us far to the westward of Cape Haitien. From the Yaque back to Monte Cristi, then straight across to Limonade Pass and up to Limonade Bord-de-Mer is about 11 leagues.

In his Journal for the same day Columbus left sailing directions for reaching Navidad from Monte Cristi, but Las Casas omitted these in his abstract "because that land is already known." Ferdinand Columbus, however, gives a digest of them in the Life of his father. You must put out to sea 2 leagues beyond Monte Cristi until you find Cabo Santo (Point Picolet), and from there the "village of Navidad is distant 5 leagues, and will

---

[75] *Journal*, January 2, 1493.

[76] "Cape of the Serpent." This is the only use of that name. Mr. Pettigrew suggests that it means the two rounded hills called Les Mamelles, on the coast just before reaching the entrance to Fort Liberté. From the sea off Limonade Pass these hills (as I have observed) look like a cape in the form of a sea-serpent with two humps, and they mark the SE end of the barrier reef. But they are over 5 leagues from Point Picolet.

[77] This is not correct; possibly Columbus was thinking of the reefs off Monte Cristi. The Seven Brothers may have seemed to be off *Cabo de Sierpe* from his angle of approach.

[78] This means to the westward, and is correct.

[79] The 1-fathom line on the chart extends $1\frac{1}{4}$ miles northward from Point Picolet, and from 3 to $4\frac{1}{2}$ miles westerly; but there is an 8-fathom sounding a league off-shore.

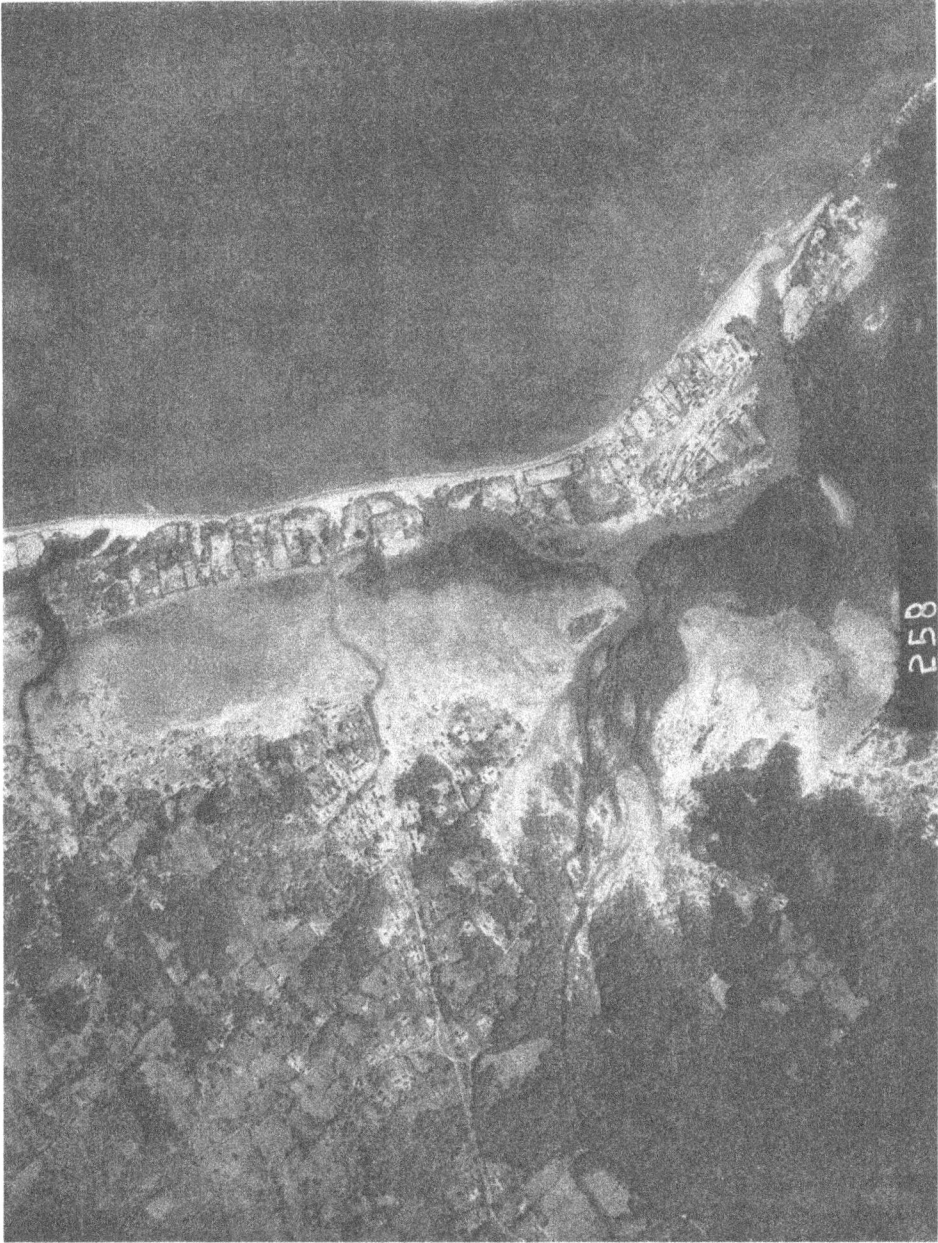

LIMONADE BORD-DE-MER FROM THE AIR.

Grande Rivière is at the left (NW) of the picture, the village is near the upper right (SE) side. The church is the long building just to the left of a group of trees beside the salt estuary. The old French fortification mound is in the partially cleared space near upper right. Dark vegetation between beach dunes and salines is mangrove.

enter by a sure channel between those shoals where he was the day before." [80]   Leagues is, of course, a mistake for Roman miles.   From a point about half-way between the end of the barrier reef and the breaking shoals that lie to the E of Cape Haitien, Limonade Bord-de-Mer is just 3¾ nautical or 5 Roman miles distant.

This concludes the evidence from Columbus's Journal.   We now have:

d. Evidence from Dr. Chanca

Dr. Chanca, the fleet physician on Columbus's Second Voyage, wrote a letter home in January or February 1494, which is our most important source of information for that voyage.[81]

The Admiral, this time with a fleet of 17 sail, approached Navidad from the eastward.

November 27, 1493.   "Navidad is 12 leagues from Monte Cristi."   This, as we have seen, is almost correct for the distance from Monte Cristi to Limonade Bord-de-Mer. The fleet anchors that night off Navidad "less than one league" from shore, and fires guns, but receives no reply.[82]

November 28.   Some of the men land, go to the spot where Navidad was, find the fort burned and levelled to the ground, and clothing strewn about.[83]

November 29.   The Admiral and the Doctor go ashore and verify what the men had seen.   Then in order to "examine a spot which lay about a league from us proper to erect a town . . . some of us went with him, observing the land from along the coast (*por la costa*)."   They reach a small village where they find many relics of the Spaniards, including the *Santa Maria's* anchor, and return to the site of Navidad.[84]

Referring to the map, it will be seen that the one place in this region where one could walk or row almost a league along the coast and find a possible site for a settlement is from Limonade Bord-de-Mer to Point Sable.[85]   That was a natural place for Columbus to examine with a view to a new settlement.   And it was on the Habitation Bellevue extending along the right bank of the Grande Rivière to the sea, that there was dug up under four feet of silt, in 1781, an iron anchor nine feet two inches long, which probably was this very one.[86]   It was exhibited at the World's Fair at Chicago in 1893, and is now in the Musée Historique of Port-au-Prince.   The design of anchors has changed so little throughout the centuries that one cannot positively identify this one as of the Columbian era; but the fact of its being found at so great a depth under the alluvial deposits of the Grande Rivière, and so near to the spot where Dr. Chanca saw the *Santa Maria's* anchor in 1493, creates the presumption that it is indeed the anchor of the flagship, which the Indians had carried off when they broke up Navidad.

[80] *Historie*, ch. 34; Caddeo ed., I, 212.

[81] For a discussion of this letter, see S. E. Morison, *The Second Voyage of Columbus* (Oxford, 1939), pp. 9–12.   I have here used the text in Navarrete, which is reprinted, with a fair translation by Cecil Jane in his *Select Documents illustrating the Voyages of Columbus*, I (Hakluyt Soc. Publ., 2d ser., LXV), 20–73.

[82] Navarrete, *Colección*, I, 212.

[83] *Id.*, p. 213.

[84] *Id.*, pp. 214–15.

[85] From Petite Anse to Cape Haitien is also a possibility, but, as we shall see, Columbus explored Cape Haitien Harbor with a view to settlement a few days later.   From Moreau de Saint-Méry's distances I believe that the Grande Rivière emptied into the bay west of Point Sable in his time.

[86] Moreau de Saint-Méry, *Description*, I, 189.   He says that it was discovered 900 toises (about 1900 yards) from the sea.

November 30. The Admiral sends a caravel in one direction to look for a suitable location for a settlement, while he and Dr. Chanca sail in the other direction and discover a harbor "very secure and of very favorable disposition of land to inhabit." But as it lay in the wrong direction from the source of gold, they decide against it. This harbor must have been that of Cape Haitien, for Columbus well knew that the Cibao, the source of gold, lay to the eastward.

On their return thence the Admiral and Dr. Chanca "as they went along the coast" meet the other caravel, which had been boarded by Guacanagari's brother. The principal Spaniards then went ashore at the cacique's village and talked with him.[87]

Here is proof that the village lay to the eastward of Navidad, and a strong suggestion that it was some distance to the eastward, at Caracol or Yaquezi.

December 1. The Admiral decides to visit Guacanagari's village, "to which place we could arrive in 3 hours, since it is less than 3 leagues distant."[88] From the anchorage off Limonade Bord-de-Mer to Yaquezi is about 7 miles (2⅓ leagues), and to Caracol a little less. They make the trip; Guacanagari accompanies them back to the ship, and returns to his village that evening after supper. He advises the Admiral that the site of his village "was unhealthy because it was very damp; and such it certainly was," affirms the Doctor.[89] We cordially agree. In the rainy season, which generally lasted in Northern Haiti until well into December, the Trou River overflows the plain at Caracol.

December 2. "We anchored in that port," off Guacanagari's village, and sent messages ashore. In the night some of the Indian girl captives escape by swimming ashore; four were taken when leaving the water after swimming "more than a good half league."[90]

December 3. This day, "because the weather was foul" (tempo contrario), "the Admiral decided it would be a good idea to go with the boats to see a harbor further up the coast at two leagues' distance,[91] to see if the land was suitable for a settlement; thither we went with all the ships' boats, leaving the ships in the harbor." They landed and inspected the harbor, evidently a steep-sided one, since they found a wounded Indian "stretched out on the mountain." The Admiral did not consider this harbor "healthy," and decided to return "up the coast[92] by which we had come from Castile, because the news of gold was in that direction."

Since Dr. Chanca uses the same general direction "up the coast" both for the harbor and the return voyage to the eastward, Navarrete identified this harbor inspected by the

[87] Navarrete, I, 215-16.

[88] Id., p. 216, foot. Nicolo Syllacio's tract of 1494, Ad sapientissimum L. M. Sforzam, an embellishment of letters to him from Guillermo Coma, a Spaniard on the Second Voyage, in most respects corroborates Dr. Chanca. Of this visit, however, Syllacio writes ad visendum regem qui ferme a mari decem milibus passuum se continebat, "to see the king who kept himself almost ten miles away from the sea." Facsimile in J. B. Thacher, Columbus, II, 234. This corroborates the archaeological evidence of Mr. Boggs (see below) to the effect that the Indian village lay some distance from the shore, but the other evidence shows it could not have been more than a mile and a half from salt water at most.

[89] Navarrete, I, 218.

[90] Id., p. 218. Peter Martyr also mentions this incident, and names the ringleader of the girls Katherine. He says that they swam "three long miles" and that three were recaptured. Eden trans. (1555), f. 8 verso.

[91] un puerto la costa arriba, fasta el cual habria dos leguas. Navarrete, I, 219.

[92] la costa arriba.

Admiral as Fort Liberté Bay. I cannot accept this identification. The mouth of Fort Liberté Bay is over 3 leagues from Caracol Pass, and the ships' boats would certainly not have gone outside in weather that kept the ships at anchor. The shores of Fort Liberté Bay are low and flat, not mountainous, although there are some small hills either side of the entrance. And this bay was discovered, as we shall see, by a fleet of the smaller barques with three hundred men under Melchior Maldonado, when they were trying to round up Guacanagari, a few days after. Cape Haitien harbor, already reconnoitred on November 30, must have been the one inspected December 3, despite Dr. Chanca's lubberly *la costa arriba*. It was a natural boat expedition for a foul day, and lay just about 2 leagues from the Limonade Bord-de-Mer anchorage, by way of Point Sable. Cape Haitien is the only harbor at anywhere near that distance which has hilly shores. And if we accept this identification of the harbor visited on December 3, Cape Haitien harbor is ruled out both as the site of Navidad and as the site of Guacanagari's village.

e. Evidence from Las Casas

Bartolomé de las Casas, who first came to Hispaniola between 1498 and 1502, viewed the site of Navidad some time before 1527, when he is supposed to have begun his *Historia de las Indias*. Therein he remarks:

"It was named the Town of the Nativity, because it was on that day that they came there, and today that harbor is still called *Navidad*, although there is no reminder that there ever has been a fort or any kind of building there, since it is overgrown with so many and such great trees (and I have seen them) as if fifty years had passed; the reason is that such is the fertility and fatness of that island that if you cut off a branch of a tree and stick it in a hole two or three palms deep, without any watering or care it will grow in two or three years to a tree little smaller than the one from which it was cut " [93]

Disappointingly vague as this statement is in respect of location, it does prove that Navidad was not on the site of Puerto Real, or any other settlement existing when Las Casas wrote, and that the anchorage was still valued as a harbor. Thus, Las Casas's description is consistent with the two earlier maps, which fairly definitely (though not indisputably) point to Limonade Bord-de-Mer as the site of Navidad, and with the later Bologna map (hereinafter described) that places Puerto Real on Fort Liberté Bay.

f. Conclusion as to Navidad and Indian Village; archaeological evidence

This concludes the contemporary evidence. Despite some discrepancy in the longer distances, which were the most difficult to measure, the evidence points unmistakably to a place for Navidad not far from Limonade Bord-de-Mer. Only from an anchorage off that place can one take a NW course to the open sea and clear the reefs; only from there can one walk or row a league along a coast unobstructed by mangrove; at the end of that league the anchor was recovered in 1781; only from Limonade Bord-de-Mer as a starting point can one find a steep-sided harbor two leagues to the westward by boats.

As to the site of Guacanagari's village the evidence is not so precise, although it points unmistakably to a site *east* of Navidad, and on Caracol Bay. Columbus's Journal for December 25 says that the village was "a good $3\frac{1}{2}$ leagues" east of Point Picolet; Caracol is $3\frac{3}{4}$ leagues, which is near enough. On the same day he records that the village

[93] *Historia de las Indias*, lib. i, cap. 61; 1927 ed., I, 285–86. Navidad is also mentioned in his *Apologética Historia*, caps. 8, 9 (*Historia*, 1927 ed., III, 431, 435), but without any clue as to its location.

THE SITE OF NAVIDAD.

Above: Fishing Fleet at Anchor off Mound of Old French Fortification.
Below: Church of St. Philomena, Limonade Bord-de-Mer.

was a league and a half from the reef where the *Santa Maria* was wrecked. Caracol is 2 leagues from the middle reef, and a little less from the eastern one. Dr. Chanca says that the village was "less than 3 leagues distant" from the fleet anchorage on the Second Voyage, which we suppose to have been off Navidad. Yaquezi is $2\frac{1}{3}$ leagues from the anchorage off Limonade Bord-de-Mer; Caracol a trifle less. Our only choice, then, is between Caracol and Yaquezi for Guacanagari's capital. All else on Caracol Bay is mangrove swamp; and the Indians were not swamp dwellers. And the next possible site for a village east of Yaquezi is on Fort Liberté Bay; but when Melchior Maldonado explored that bay after December 3, 1493, he encountered a cacique "with a frownynge countenaunce and a grymme looke, with a hundreth men folowynge hym," who belonged to a different tribe.[94]

Hence I conclude that Guacanagari's residence was within a mile of the short stretch of open shore at Caracol, secure from spring tides but handy to his canoes.

Limonade Bord-de-Mer [95] today is a small fishing village situated on a line of low sand dunes behind the beach. About twenty-five poor houses are clustered around a single French colonial church dedicated to St. Philomena, who is held in great reverence by the fishermen of that region. The church, not more than 125 metres from the sea, is a landmark for vessels entering by Limonade Pass. Across a little salt estuary—probably a former mouth of the Fossé, and represented by the narrow slit on the La Cosa Map— the local fishing fleet is moored close to shore. There, as one of the fishermen explained, rather than off the village or the beach, they find the best holding ground and protection for small craft. We asked ourselves, if Columbus found the best holding ground, the best protection, and the dryest site in this whole region, on the nearest mainland to the wreck, would he have sought further? The one unsuitable factor about this particular site, or any in the neighborhood of Limonade Bord-de-Mer, is the lack of fresh water. In colonial days the people here had to haul their drinking water from the Grande Rivière.[96] The present fisher folk have sunk a couple of wells behind the church, but they say that the water is brackish, suitable only for washing and cooking, not for drinking. It is possible however that the present salt estuary was, in 1492, the mouth of the river Fossé, and that fresh water was then obtainable by sending boats a short distance up-stream.

East of the salt estuary the line of sand dunes continues, but greatly diminished; and within about a quarter mile from the church the mangroves come right up to the dunes. Finally, about one mile from the church in this same direction begins the mangrove swamp, which reaches around into Caracol Bay. Consequently, the possibilities of a site of Navidad east of the salt estuary are limited to a space along shore within a quarter mile of the church.

Within this space, facing the shore, and exactly abreast of the modern fishing-boat anchorage, is a mound shaped like a flattened and inverted letter U, flat on top and rising about 15 feet above the surrounding country. The regularity of the mound left no doubt that it was artificial, and the history of the locality suggested that it was one of the three

---

[94] Peter Martyr d'Anghiera, *The Decades of the newe worlde of West India* (Richard Eden, trans., London, 1555), p. 9 verso.

[95] The village is called *L'Embarcadère de Limonade* by Moreau de Saint-Méry. A more considerable place in his day than in ours, three sailing packets plied daily back and forth to Cape Haitien. *Description*, I, 209–12.

[96] *Id.*, I, 211.

redoubts built by the French here in order to protect *l'embarcadère*.[97]   A small powder house of brick and coral stone, situated between the wings of this mound, is undoubtedly a survival of the French fortifications.   Columbus certainly had no time to erect such a mound between Christmas, 1492, and New Year's, 1493.   But the possibility of these French fortifications having been erected on top of the remains of Navidad seemed so strong, that the mound itself seemed worth excavating.

Accordingly, in June 1939, Mr. Stanley H. Boggs, of the Peabody Museum of Archaeology of Harvard University, undertook a test excavation of this mound and other nearby sites, with funds kindly furnished by one of my classmates.[98]   Six test excavations of the mound were made by local workmen under Mr. Boggs' directions as follows:

1. Trench 15 metres long N to S, 2 m. wide, 1 to 2.10 m. deep, in center of mound.

2. Trench 2.5 m. long E to W, 2 m. wide, on NW corner and outer slope of mound, extending from base to crest.

3. Trench parallel to and 3.8 m. N of Test 2, depth 1.2 m.

4. Trench parallel to Tests 2 and 3, and between them, with same dimensions but from 1.55 m. to 1.8 m. deep.

5. Trench 2.5 m. long E to W, 2 m. wide, on NE corner and outer slope of mound, varying from 1.85 m. to 2.1 m. deep.

6. Two small pits 1 m. in diameter, one 3 m. W of powder house, and another at S end of E wing of mound.

These six test excavations covered all parts of the mound in such fashion that, if any Columbian remains were below the French fortifications, there was very little likelihood of their escaping pick and shovel, or the expert eye of Mr. Boggs.   Three alternate layers of sand-coral and ashes-earth were found in Test 2.   In all three specimens were found, consisting of corroded bits of iron (mainly nails), some pottery reputedly French colonial, and a few animal bones.   In the ashes were found bits of charred bayahonde wood, such as would have been used for cooking fires.   Bits of brick were found only 20 cm. above the water level.   There was no sequence in the occurrence of the specimens.

Tests 1 and 3 yielded no specimens.   Test 4 turned out artifacts similar to those of Test 2.   Test 5 yielded a few pieces of glass and of rusty iron; stratification here was extremely vague.   Test 6 yielded only sherds of modern pottery and bits of brick.   None of these specimens are demonstrably of Columbian or even Spanish origin; although there is a possibility that some of the nails, if analyzed by a competent metallurgist, might prove to be of 15th century iron.   The paucity of charcoal at the bottom of the pits satisfied Mr. Boggs that this was not the site of a razed wooden structure such as Navidad; but he does not exclude the possibility that the charred embers of the fort might have floated away in a flood or hurricane, or sunk below water level.

Having drawn a blank from the mound, Mr. Boggs tried test excavations in other places around Limonade Bord-de-Mer that suggested a possible fort site to the experienced archaeologist's eye.   Two of these were east of the estuary, near the mound; fifteen others

[97] Moreau de Saint-Méry, *Description*, I, 210–12.   The place was somewhat fortified before 1717, but the existing works date from the latter part of the century.   During the American War of Independence, when the French garrison of Cape Haitien accompanied the fleet of De Grasse to Yorktown, these redoubts were garrisoned by Spanish troops from Santo Domingo.

[98] Mr. Boggs proposes to publish a detailed report of his excavations, but has consented that I incorporate his preliminary report to me in this article.

were in or near the village.   Most of them yielded no specimens; in a few, various artifacts of French colonial or more recent Haitien origin were found.

The result of these excavations, then, is negative and inconclusive.   No evidence of Navidad was found; but the possibility that no remains are left is so strong that a negative excavation proves nothing.   The period of Navidad's human occupancy was only a few months; and when the Indians destroyed the garrison they would certainly have carried off every bit of metal they could lay hands on.   All vestiges of clothing would long since have disappeared, and although bits of the burnt wood may still be *in situ*, it is much more likely that they were carried off in the first hurricane of 1493.   Mr. Boggs concludes "no amount of excavation near Bord-de-Mer would guarantee the discovery of Navidad unless more definite evidence of its location is procured."

So, in spite of these negative results of excavation, I still believe that Navidad must have been within half a mile of the church at Limonade Bord-de-Mer.   Anything further to the eastward takes us into the midst of the mangrove swamp; and the further one goes to the westward along the beach, the worse becomes the anchorage and the protection.

What of the site of Guacanagari's village?   As we have seen, the literary evidence points to a place at or near Caracol or Yaquezi.   At Caracol there are two *embarcadères* or landing places, with ruins of what must have been large port buildings and defenses, which Moreau de Saint-Méry says were begun in 1713.[99]   South of the landings is a flat, sandy strip about a third of a mile wide, which one crosses to visit the village of Caracol. Possibly it was on this *playa* that Columbus indulged in archery, to the admiration of Guacanagari.   Near the Garde de Haïti headquarters at Caracol village are six small shell-heaps, which Mr. Boggs excavated, but found no artifacts older than French colonial. He observes that, if the original village had been here, its remains would probably have been leached away by the periodical floods of the Trou River.

A much more promising site than that of Caracol village was found by Mr. Boggs on the plantation of the Haitian Agricultural Corporation managed by Mr. Crooks, about a mile inland from Caracol Embarcadère on elevated ground, and not far from the Trou River.   Here was an extensive Indian village site with six large mounds.   None could be thoroughly excavated because they were planted with sisal, but Mr. Boggs dug a number of trenches and pits and found many Indian artifacts.   None of these could be definitely assigned to pre-Columbian days.

A third possible site is at Yaquezi (pronounced "Jacksie"), half a league to the eastward along Caracol Bay.[100]   Here is the same wide strip of flat sand between the landing place and the village; here also are the ruins of French port buildings and defensive works (begun in 1762), and indications of Indian kitchen-middens.   One cannot see the bay from the landing-place, which is approached by a curved channel through the mangrove swamp; and as Carib enemies approached by sea, the mangrove screen would possibly have recommended this site to the Arawaks.   About a mile south of Yaquezi, on the road to Trou village, Mr. Boggs noted a promising village site of some 30 shallow mounds covering at least half a square mile.   Many of the mounds have Haitian houses on top of them, and for this and other reasons it was not practical to excavate the site.

[99] *Description*, I, 162.

[100] That part of Caracol Bay east of the short mangrove-covered points between Caracol and Yaquezi is called Yaquezi Bay by Moreau de Saint-Méry.

LIMONADE BEACH.

Above: View taken half a mile West of Village.
Below: Mouth of Grande Rivière, showing Pointe Picolet, Cape Haitien.

This and the Crooks site, each about a mile from the shore, Mr. Boggs regards as very probable sites of Guacanagari's village.

g. False Sites: Puerto Real

There remains to examine various hypotheses that have been earlier brought forward for the site of Navidad, and to state the reasons which have caused me to reject them.[101]

The earliest historian to attempt to solve the question was Antonio de Herrera. In his *Descripción de las Indias Ocidentales* written about the year 1600, Herrera says that the Spanish town of Puerto Real was founded "where the Admiral . . . went ashore, the first time that he came to this island," and for that reason was given these arms: *azure wavy a ship or.*[102]   But where was Puerto Real?   One of the fortified towns founded under Ovando in 1503 in order to pacify the Indians, it was abandoned before 1606, and that was seventy years before the French occupied this part of Hispaniola.[103]   The maps in Charlevoix' *Histoire de L'Isle Espagnole ou de S. Domingue* (1730) place Puerto Real in Manzanillo Bay.   Moreau de Saint-Méry thought it was somewhere on Caracol Bay.[104] But in Peter Martyr's *Decades* there is emphatic evidence that Puerto Real was the original name of Fort Liberté Bay, which was discovered and so named in December 1493 by Melchior Maldonado, an officer of Columbus's fleet from whom Martyr obtained his information.

On December 2, 1493, when Columbus was exploring the region in the course of trying to discover what had become of the Navidad garrison, he anchored off Guacanagari's village.   That night, the young Indian girls who had been captured in Guadeloupe slipped overboard and swam ashore.   The next day, when the Admiral sent a boat ashore, Guacanagari and his followers were found to have absconded, leaving their village deserted. As Richard Eden relates the incident in his quaint translation of Peter Martyr:

"Wherfore the Admirall sente forthe an armye of three hundrethe men, ouer the which he appoynted one *Melchior* [105] to be capitayne, wylling hym to make diligent searche to fynde owte *Guaccanarillus*.   *Melchior* therfore with the smauleste vessels enteringe into the countreye by the ryuers and scouringe the shores, chaunced into certen croked goulfes defended with. v lyttle and stiepe hilles, supposinge that it had byn the mouth of sum greate ryuer.   He founde here also a verye commodious and safe hauen, and therefore named it *Portus Regalis*.   They saye that the enteraunce of this is so crooked and bendinge, that after the shippes are once within the same, whether they turne them to the lefte hand, or to the ryght, they can not perceaue where they came in vntyll they returne to the mouth of the ryuer: Although it be there so brode that three of the byggeste vessels may sayle together on a front.   The sharpe and high hilles on the one

---

[101] I have not thought it worth while to demolish the snap judgments contrary to the evidence made by various secondary writers who have not really studied the question.   For instance, the map in Rinaldo Caddeo's edition of Ferdinand Columbus's *Historie*, I, 201, locates the shipwreck on a reef only a mile from Point Picolet, in spite of the Admiral having written (and his son having repeated) that the *Santa Maria* had already sailed a league east of that point at 11 p.m. on the evening of the shipwreck, and places Navidad on a hill in Cape Haitien Harbor.   And the map in Markham's *Columbus*, reprinted in Filson Young's *Columbus* (3d ed., 1912), p. 236, places Navidad west of "C. Santo."

[102] Incorporated in his *Historia General de los Hechos de los Castellanos*, dec. I, lib. vii, ch. 2 (1730 ed., I, 179).   Although most of Herrera is lifted from Las Casas, I cannot find this statement there.   These arms are depicted on the D'Anville map of Haiti in Charlevoix, *Histoire de Saint Domingue* (1730), I, 236.

[103] Moreau de Saint-Méry, *Description*, I, 163–64, 181.

[104] *Id.*, p. 163.

[105] Melchior Maldonado.

CARACOL FROM THE AIR.

The white area at the left (W) of the shore line is Petit Caracol; the one near the middle, Grand Caracol. The light area is probably the *playa* where Columbus put on an archery exhibition for Guacanagari. Caracol village can be seen just south of this inland beach. The most probable site of Guacanagari's village is about a mile to the SW.

syde and on the other, so brake the wynde, that they were vncerten howe to rule theyr sayles. In the myddle gulfe of the ryuer, there is a promontorie or point of the lande with a pleasaunte groue full of Popingayes and other byrdes which breede therein and singe verye sweetlye. They perceaued also that two ryuers of no smaule largenes fell into the hauen." [106]

This description is a remarkably exact one of Fort Liberté Bay, except for the five "lyttle and stiepe hills," which seem to be an exaggeration of high banks. Obviously the town of Puerto Real, founded in 1503, took its name from the harbor discovered and so called ten years earlier. The nautical arms granted to it may have been a reminiscence of the neighboring Navidad, but more likely referred to the excellence of the harbor.

This location of Puerto Real is confirmed by a map of Hispaniola by an unknown artist dated 1516, and now in the Royal Library at Bologna.[107] Herein the indentation of the coast between Cape Haitien and Monte Cristi is grossly exaggerated, but there is much detail. Navidad is not shown, for it was long since deserted; but on a small bay, with the unmistakable shape of Fort Liberté, located west of the river *Dahabon*, the modern Massacre,[108] is the name *Portus Regalis*, and on the nearest blank space, a fancy picture of a fortified town. Thus, there is no doubt that Puerto Real was on Fort Liberté Bay, and that Navidad was not on the site of Puerto Real.

h. False Sites: Cape Haitien Harbor and Petite Anse

Local opinion names Fort St. Michel [109] as the site of Navidad, and places the village at Petite Anse. Fort St. Michel is on an outcrop of solid rock rising to a height of about 50 feet above mangrove swamps and salines, next the airplane landing field, a few hundred yards from the beach at the head of Cape Haitien Harbor, and about five-eighths of a nautical mile from the Embarcadère de la Petite Anse. At Cape Haitien it is the general belief that Fort St. Michel was the site of Navidad, and as such it has been often exhibited to tourists, who are even told that the remains of a nineteenth-century stone fortification there [110] were constructed by the Admiral.

There are several insuperable objections to Fort St. Michel when we compare its location with the documentary evidence of Navidad. It is impossible to sail a NW course from the harbor in front of it; and no juggling with compass deviation, a favorite sport of library navigators, can overcome that obstacle; for the Admiral's map shows that he was correctly oriented. Fort St. Michel is 4 miles from the nearest possible reef where

---

[106] Peter Martyr d'Anghiera, *The Decades of the newe worlde of West India* (Richard Eden trans., London, 1555), ff. 8 recto–9 verso. Cf. Syllacio's tract, in Thacher, *Columbus*, II, 236. He says that this "very safe harbor," 15 Roman miles from Navidad, was called *regalis*, because second to none in the world.

[107] *Es Map mas antiguo de la Isla de Santo Domingo* (Carlo Frati ed., Florence, 1929; extrait de *La Bibliofilia*, t. XXIII). I am indebted to M. Edmond Mangonès of Port-au-Prince for calling my attention to this map. See outline sketch of this section on page 262.

[108] Moreau de Saint-Méry, *Description*, 1797–98 ed., I, 108. On this Bologna map the *Jaq Fl[umen]* is the Trou River, which was called the Yaquezi or Jaquesi in Moreau de Saint-Méry's day; next to the westward, around a much flattened Caracol Point, is *Maimon Fl[umen]*, the Grande Rivière; and next, the Rivière Haut du Cap.

[109] Called Fort Vilton on the U. S. charts.

[110] Moreau de Saint-Méry refers to the hill only as *un petit bout de montagne*, and does not include the fort in his exact description of the fortifications of Cape Haitien. It was probably built at the time of the Napoleonic wars.

CARACOL AND YAQUEZI.
Above: Embarcadère de Grand Caracol.
Below: Salines of Yaquezi, looking South.

the *Santa Maria* could have been wrecked, by a tortuous boat channel through numerous coral heads; it is less than 3 miles from Point Picolet.   In addition, it is three miles *west* of the point where the anchor was found; it would have been impossible for the Spaniards to have sunk a well in solid rock, difficult for them even to have built the fort there in a week's time, and useless to palisade.   Moreover, it is very likely that in 1492 this hill was inaccessible from the sea by reason of mangrove swamps, which were filled up by the French after the region became thickly populated and the land valuable.

Mr. T. S. Heneken, a resident of Cape Haitien to whom Washington Irving applied for information almost a century ago, declared that *Guarico*, the Spanish name for Petite Anse, was derived from Guacanagari, and proves that his village was there situated.[111] He assumes that Navidad was located some distance up the Rivière Haut du Cap which flows into Cape Haitien Harbor just south of the town.[112]   The same objections apply to this hypothesis as to the previous one.   *Guarico* may well have been derived from Guacanagari, whose dominion extended over the entire northern part of Haiti; he may even have shifted his headquarters thither between 1494 and the next visit of the Spaniards. Columbus says that Guacanagari's village lay a good 3½ leagues *east* of Point Picolet. Petite Anse lies one league *south* of Point Picolet.   Dr. Chanca says that the village lay 3 leagues east of Navidad by sea.   Petite Anse is less than a mile from Fort St. Michel by land.

i. Moreau de Saint-Méry's Site

Moreau de Saint-Méry had definite ideas as to the sites of Navidad and of the cacique's village; and he is so accurate an observer of his beloved Saint-Domingue that his theory merits a respectful examination.   But we must remember that he had no authority older than Hererra's *Historia General* (1601); he knew neither Las Casas's précis of Columbus's Journal, nor his *Historia de las Indias.*

Speaking of Caracol, Moreau de Saint-Méry writes:

"Mais ce qui est encore plus glorieux pour Caracol, c'est que son port est celui de la Nativité, ainsi nommé par Christophe Colomb, qui y entra le jour de Noël 1492.   En rapprochant tout ce qu'il y a de descriptif dans les premiers historiens du Nouveau-Monde, il n'est guères possible de douter de ce fait; surtout quand on remarque que le chef-lieu du royaume de *Guacanaric* était sur une pointe, à l'extrémité de la Véga-Réal,[113] et conséquemment vers le point où est maintenant l'embarcadère de la Petite-Anse, au Quartier-Morin,[114] et qu'il est dit que Colomb partant de la Nativité, fit de l'eau au Nord-Ouest, puis sortit en remarquant bien l'entrée pour la reconnaître, que son lit était noyé et qu'on n'y trouvait point de pierres pour bâtir; circonstances qui semblent bien désigner la rivière de Caracol ou de Jacquezy, une entrée aussi difficile que celle de Caracol. . . .[115]   Ce fut près du port de la Nativité que fut construite la tour que l'on appella la forteresse de la Nativité, et où Colomb laissa quelques Espagnols, qu'il trouva massacrés à son second voyage. . . .[116]

[111] Moreau de Saint-Méry (*Description*, 1797 ed., I, 296–97) says that *Guarico* was the Spanish name of Cape Haitien, not of Petite Anse.

[112] Washington Irving, *Life and Voyages of Christopher Columbus* (Vol. 3, *Works of Irving*, N. Y., 1849), I, note at end of ch. XI in book iv (pp. 226–27).

[113] The plain between mountains and sea.

[114] This seems to identify Moreau's Petite Anse as the place of that name today, although General Nemours of Port-au-Prince informs me that in the eighteenth century the name was applied to a cove further east.

[115] Apparently he means Caracol Pass.   This has never been anything more than a boat passage in historic times, and I doubt if Columbus learned about it.

[116] *Description*, I, 163–64.

Although Caracol is accorded the glory of having the harbor of Navidad, Moreau de Saint-Méry believed that the fortress lay some distance inland from Limonade up the Fossé river. In the following passage he attempts to identify the exact site:

"Sur la rive Orientale du Fossé, à environ une lieue de son embouchure actuelle et dans la partie la plus élevée de la savane de Limonade, on a trouvé sur un terrain, dépendant à présent de l'habitation Montholon, à deux ou trois cens toises [117] des bâtimens de cette habitation, les fondemens d'un fort, considéré comme celui de *la Nativité*, construit au mois de Janvier 1493, par Colomb. Ces fondemens étaient de la pierre aimantaire du morne à Békly.[118] Ils ont été démolis pour construire les bâtimens de l'habitation Destouches à qui ce local appartenait alors." [119]

Mr. Pettigrew and I, by following Moreau de Saint-Méry's accurate description, managed to identify and visit the site of this "fort." It is just east of the main road and of the now dry bed of the Fossé.[120] A mound artificially levelled arises about fifteen or twenty feet above the general level of the plain; the ruins of the Habitation Montholon are visible about 600 yards to the northwestward. Some bricks of French or Spanish make were lying about on the surface.

There are several reasons which lead us to reject this as the site of Navidad. During the week that Columbus had to construct the fort, he hardly had time to procure rock from the Morne Békly almost two miles distant, much less to instruct his volunteer Indian helpers, as Moreau de Saint-Méry suggests, in the making and baking of brick.[121] Even if the Fossé was navigable by boats in colonial times up to the present main road, as Moreau de Saint-Méry says it was,[122] there is no reason to suppose that Columbus would have built his fort two or three miles up a river, rather than on the coast. Moreau supports his hypothesis by Herrera, whom he misunderstood to say that Columbus, on landing in November, 1493, went up a river in boats in order to visit the site of the fort. There is no such statement or even suggestion in Herrera.[123] Why should Columbus have gone to the additional labor of boating materials from the wreck up a river to construct a fort, when an obvious site offered itself on the bay? All other forts constructed by the early

[117] One toise = 6.4 feet, or 6 feet 4¾ inches.

[118] The Morne à Békly is a prominent rocky hill that rises less than 100 feet from the plain, about 2 miles as the crow flies from this site. It was the common source of building stone for the French habitations. This stone contains magnetic iron, and the morne has given much trouble to surveyors by deflecting their compasses.

[119] *Description*, I, 206.

[120] A few yards along the road, in the direction of Cape Haitien, is a sharp corner where there is a smithy and a few houses. A few yards further a large tree marks the beginning of the old road (now impassable for cars) to Limonade Bord-de-Mer.

[121] Mrs. Pettigrew informs me that no fragments of bricks have been found in the numerous Indian mounds of this region that she and others have excavated.

[122] *Description*, I, 189. The Habitation Walsh that he there mentions is on the south and west side of the road.

[123] The passage from Herrera's *Historia General*, dec. i, lib. ii, ch. 9, as correctly translated by Capt. John Stevens (London, 1740), I, 113, is: "The next *Monday* all the Fleet enter'd the Port, the Admiral saw the Fort burnt down, whence he concluded, that all the Christians were dead, which troubled him very much, ... The next Day he went ashore very melancholy, finding no Body to enquire of. Some Things belonging to the *Spaniards* were found, the Sight whereof was grievous. He went up the River with the Boats, and in the mean Time gave Orders for cleansing a Well, he had before made in the Fort. . . . Near the Fort they discover'd seven or eight Men bury'd, and others farther off. . . ." It is clear from this that the Admiral went up the river after he had examined the fort, and then returned.

Spanish and Portuguese discoverers were on or very near the sea, in order to command an approach by hostile canoes or vessels.

Our ingenious searcher supports his conjectural site by the following discovery:

"Une tradition constante appuie le fait, et l'on a encore trouvé en 1784, lors d'une fouille pour le canal du moulin de l'habitation Montholon, non loin du château, une espèce de tombeau où il y avait vingt-cinq cadavres qui n'appartenaient point à des Indiens, puisque leurs têtes n'étaient point applaties. Ces corps dont on distinguait la charpente osseuse étaient dans la même direction et parallèlement disposés, usage que l'on sait qu'avaient les Indiens pour leurs morts: c'étaient donc des Espagnols enterrés pas des Indiens. Enfin l'on a trouvé dans le même lieu des fourchettes de fer bien rouillées et des pièces de cuivre." [124]

Moreau de Saint-Méry owned two of these coins, and gives a sketch of them; one had the date 1476. From his sketch it is clear that they were copper coins of the reign of Ferdinand and Isabella, and the date 1476 is quite possible. The presence of table forks, however, seems to date this "tomb" considerably later than Columbus's time; for forks were not invented until after his death, and did not come into general use even among the upper classes until the seventeenth century. No doubt the bodies were those of Spaniards, victims of some later battle, massacre or epidemic; perhaps they were some of those killed in the Battle of Limonade with the French that took place near that very spot in 1691.[125]

Moreau de Saint-Méry's attempt to identify Navidad was the only one previous to ours, made by a person who had actually studied the terrain. Not having access to Las Casas's *Historia* or Columbus's Journal, he was misled by a misunderstanding of Herrera into seeking an inland site for the Spanish fort. Probably it will never be possible to prove the site, because even a thorough excavation of the correct one may not find any relics of so ephemeral a settlement. The clothes, for which the natives had no use, and which they left strewn about, will have rotted completely away; and it is not likely that the Indians failed to carry off any bit of metal.

7. From Monte Cristi to Samaná

Beyond Navidad I have not been able to check up on Columbus's course by personal inspection, except for brief visits to Monte Cristi in 1938 and 1939, and to Samaná Bay in 1938. I shall not, therefore, attempt more than a brief sketch of the Admiral's First Voyage along the north coast of the present Dominican Republic.

Sailing from Navidad on January 4, 1493, and coming out to sea, the Admiral directed his course east to "a very high mountain which seems to be an island but is not one, because it is connected with very low land, and which has the shape of a very fine tent [126] to which he gave the name *Monte Cristi*, which lay due east of Cabo Santo, distant 18 leagues."

His description of Monte Cristi is exact, although its distance from Cape Haitien is exaggerated and the bearing is a point out. Sailing to the eastward this high cape first appears as a high island of a striking yellowish color, resembling a tent with a ridgepole. Later Spaniards likened it to a barn, and the peninsula is still called La Granja. One does not see its connection with the shore until about half the way from Cape Haitien.

[124] *Description*, I, 208.
[125] *Description*, I, 182–83.
[126] *alfaneque*, a large Moorish tent or pavilion.

Owing to a light wind, the *Niña* did not make Monte Cristi by nightfall on January 4. She discovered the islets now known as the Seven Brothers, and anchored, says Columbus, in 19 fathoms of water, 6 leagues from the mountain. It is true that a 17-fathom spot is marked on the modern chart about 5½ leagues W of the mountain; but from the courses given next day I rather think that his anchorage was near the 18-fathom sounding about 3½ leagues NW of the mountain. In any case it was far out to sea, on the edge of the bank.

On January 5 the Admiral sailed in toward Monte Cristi, and anchored in a natural harbor between the mountain and Isla Cabra, where the modern lighthouse is located. Next day, when beating to windward east of Monte Cristi, Columbus met the *Pinta* running free; Martin Alonso Pinzón came aboard the new flagship, and doubtless high words passed between him and the Admiral for what the latter calls his "insolence and disloyalty" in parting company off Cuba. The two ships returned that night to the *Niña's* former anchorage inside Isla Cabra.

After two more days at Monte Cristi, one of which was spent in exploring the Rio Yaque, the *Niña* and the *Pinta* made sail at midnight January 8, and in 15 hours covered about 35 miles, anchoring under the lee of a cape that Columbus named *Punta Roja*, either Cape Isabela or Punta Rucia. It was at Isabela that he founded a settlement just a year later. On the 10th he made only 3 leagues to windward, anchoring in a harbor in which the *Pinta* had already spent 16 days. Martin Alonso had already named it after himself, but the Admiral renamed it *Rio de Gracia*. It is probably identical with the modern Puerto Blanco.

On the night of the 11th the fleet hove to off a mountain that Columbus named *Monte de Plata* because (said Las Casas) its summit is always covered with silver clouds. The modern Puerto Plata preserves the name. During the dawn watch on January 12 the fleet filled away with a fresh land breeze, which fortunately shifted to NNW. The *Niña* and *Pinta* went boiling along; and by sunset, when they came to an anchor in 12 fathoms inside the little island at the mouth of Samaná Bay, the caravels had covered a good hundred miles, averaging better than 8 knots.

On Sunday, January 13, at Samaná, the Spaniards had their first contact with Indians who showed fight. One of them, very ill-favored in countenance, was induced to come aboard. Shortly after, the Admiral sent a boat ashore to a point where over fifty Indians were assembled. The Spaniards purchased two of their bows and arrows, but the Indians refused to sell more; and, when the Spaniards pressed them, made a motion to pick up their weapons and attack. At once the Spaniards fell upon them with swords, and put them to flight. The Admiral named the bay *Golfo de las Flechas;* and a point just north of the Island, where the fleet anchored, is still called *Las Flechas* (the arrows) in memory of this, the first brush between Indians and white men.

From this point on January 16 the two caravels took their departure for Spain.

www.ingramcontent.com/pod-product-compliance
Lightning Source LLC
Chambersburg PA
CBHW081333190326
41458CB00018B/5984